Development of
CMOS-MEMS/NEMS Devices

Development of CMOS-MEMS/NEMS Devices

Special Issue Editors

Jaume Verd
Jaume Segura

MDPI • Basel • Beijing • Wuhan • Barcelona • Belgrade

MDPI

Special Issue Editors
Jaume Verd
University of the Balearic Islands
Spain

Jaume Segura
University of the Balearic Islands
Spain

Editorial Office
MDPI
St. Alban-Anlage 66
4052 Basel, Switzerland

This is a reprint of articles from the Special Issue published online in the open access journal *Micromachines* (ISSN 2072-666X) from 2018 to 2019 (available at: https://www.mdpi.com/journal/micromachines/special_issues/Development_CMOS_MEMS_NEMS_Devices)

For citation purposes, cite each article independently as indicated on the article page online and as indicated below:

LastName, A.A.; LastName, B.B.; LastName, C.C. Article Title. *Journal Name* **Year**, *Article Number*, Page Range.

ISBN 978-3-03921-068-8 (Pbk)
ISBN 978-3-03921-069-5 (PDF)

Contents

About the Special Issue Editors

Jaume Verd received the B.S. degree in telecommunication engineering from the Polytechnic University of Catalonia (UPC), Barcelona, in 1997 and the M.Sc. and Ph.D. degrees in electronic engineering from the Autonomous University of Barcelona (UAB) in 2003 and 2008 respectively. Since 2006, he has been with the Electronic Systems Group, University of the Balearic Islands where he became an Associate Professor in electronic technology in 2011. His research is actually focusing on the exploitation of CMOS-M/NEMS resonators features in the development of compact systems-on-chip devices for sensing and RF applications.

Jaume Segura received the M.S. degree in Physics from the Balearic Islands University (UIB), Palma, in 1989 and the Ph.D. degrees in electronic engineering from the Polytechnic University of Catalonia (UPC), Barcelona, in 1992. Since 1994, he has been with the Electronic Systems Group, University of the Balearic Islands where he became a Full Professor in electronic technology in 2007. His research is actually focusing on the exploitation of CMOS-M/NEMS resonators.

micromachines

MDPI

Editorial

Editorial for the Special Issue on Development of CMOS-MEMS/NEMS Devices

Jaume Verd and Jaume Segura

Electronic Systems Group (GSE), University of the Balearic Islands, E-07122 Palma (Illes Balears), Spain; jaume.verd@uib.eu (J.V.); jaume.segura@uib.es (J.S.)

Received: 9 April 2019; Accepted: 16 April 2019; Published: 24 April 2019

Micro and nanoelectromechanical system (M/NEMS) devices constitute key technological building blocks to enable increased additional functionalities within integrated circuits (ICs) in the More-Than-Moore era, as described in the International Technology Roadmap for Semiconductors. The CMOS ICs and M/NEMS dies can be combined in the same package (SiP) or integrated within a single chip (SoC). In the SoC approach, the M/NEMS devices are monolithically integrated together with CMOS circuitry, allowing the development of compact and low-cost CMOS-M/NEMS devices for multiple applications (physical sensors, chemical sensors, biosensors, actuators, energy actuators, filters, mechanical relays, and others). On-chip CMOS electronics integration can overcome limitations related to the extremely low-level signals in sub-micrometer and nanometer scale electromechanical transducers, enabling novel breakthrough applications. In addition, nanoelectromechanical relays have been recently proposed for mechanical logic processing and other applications in CMOS–NEM hybrid circuits spreading the More-Than-Moore approach.

This Special Issue includes 11 papers dealing with the use of CMOS-M/NEMS devices not only in the field of sensing applications (infrared sensors, accelerometers, pressure sensors, magnetic field sensors, mass sensors) but also as clock references and integrated mechanical relays. The issue covers a wide range of topics inherent to these multidisciplinary devices related to fabrication technology, mechanical and functional characterization and interfacing with CMOS electronics design.

In particular, Göktas [1] analyzes theoretically and via FEM simulations the potential of using micromachined beam structures as ultra-sensitive CMOS-MEMS temperature sensors for infrared (IR) sensing applications. In the same topic, from a more experimental perspective, Duraffourg et al. [2] report on the fabrication and characterization of a dense array of nanoresonators, with a cross-section of 250 nm × 30 nm, whose resonant frequency changes with the incident IR-radiation, allowing temperature sensitivities down to 20 mK. The work by Miguel et al. [3] outlines a novel characterization method to determine the maximum deflection of the flexible top plate of a capacitive MEMS pressure sensor based on using an atomic force microscope in contact mode. The work by Lin and Dai [4] proposes a micromagnetic field sensor based on a magnetotransistor and four hall elements with the advantage of not requiring post-CMOS processing. The work by Liu et al. [5] reviews the sensing mechanisms, design, and operation of miniaturized MEMS gas sensors focusing on the monolithic CMOS–MEMS approaches. The work by Li et al. [6] proposes a high-precision miniaturized three-axis digital tunneling magnetic resistance-type sensor with a background noise of 150 $pT/Hz^{1/2}$ at a modulation frequency of 5 kHz using an interface circuitry designed on a standard CMOS 0.35 μm technology. In the work of Perelló-Roig et al. [7], the design, fabrication, and electrical characterization of an electrostatically actuated and capacitive sensed 2-MHz plate resonator structure that exhibits a predicted mass sensitivity of ~250 $pg \cdot cm^{-2} \cdot Hz^{-1}$ is presented. The work of Riverola et al. [8] presents a tungsten seesaw torsional relay monolithically integrated in a standard 0.35 μm CMOS technology capable of a double hysteretic switching cycle, providing compactness for mechanical logic processing. Chan Jo and Young Choi [9] present a novel encapsulation method of NEM memory switches based on alumina passivation layers being fully compatible with the CMOS baseline process that allows locating

NEM memory switches in any place, making circuit design more volume-efficient. Li et al. [10], reports on a high-order ΣΔ modulator circuit fabricated in a standard 0.35 μm CMOS process acting as a low-noise digital interface circuit for high-Q MEMS accelerometers. Finally, the work of Islam et al. [11] reports a real-time temperature compensation technique to improve the long-term stability of a ~26.8 kHz self-sustained MEMS oscillator that integrates a single-crystal silicon-on-insulator (SOI) resonator with a programmable and reconfigurable single-chip CMOS sustaining amplifier.

We would like to warmly thank all the authors for publishing their works in this SI and specially to all the reviewers for dedicating their time and for helping to improve the quality of the submitted papers.

References

1. Göktaş, H. Towards an Ultra-Sensitive Temperature Sensor for Uncooled Infrared Sensing in CMOS–MEMS Technology. *Micromachines* **2019**, *10*, 108. [CrossRef] [PubMed]
2. Duraffourg, L.; Laurent, L.; Moulet, J.-S.; Arcamone, J.; Yon, J.-J. Array of Resonant Electromechanical Nanosystems: A Technological Breakthrough for Uncooled Infrared Imaging. *Micromachines* **2018**, *9*, 401. [CrossRef] [PubMed]
3. Miguel, J.A.; Lechuga, Y.; Martinez, M. AFM-Based Characterization Method of Capacitive MEMS Pressure Sensors for Cardiological Applications. *Micromachines* **2018**, *9*, 342. [CrossRef] [PubMed]
4. Lin, Y.-N.; Dai, C.-L. Micro Magnetic Field Sensors Manufactured Using a Standard 0.18-μm CMOS Process. *Micromachines* **2018**, *9*, 393. [CrossRef] [PubMed]
5. Liu, H.; Zhang, L.; Li, K.H.H.; Tan, O.K. Microhotplates for Metal Oxide Semiconductor Gas Sensor Applications—Towards the CMOS-MEMS Monolithic Approach. *Micromachines* **2018**, *9*, 557. [CrossRef] [PubMed]
6. Li, X.; Hu, J.; Chen, W.; Yin, L.; Liu, X. A Novel High-Precision Digital Tunneling Magnetic Resistance-Type Sensor for the Nanosatellites' Space Application. *Micromachines* **2018**, *9*, 121. [CrossRef] [PubMed]
7. Perelló-Roig, R.; Verd, J.; Barceló, J.; Bota, S.; Segura, J. A 0.35-μm CMOS-MEMS Oscillator for High-Resolution Distributed Mass Detection. *Micromachines* **2018**, *9*, 484. [CrossRef] [PubMed]
8. Riverola, M.; Torres, F.; Uranga, A.; Barniol, N. High Performance Seesaw Torsional CMOS-MEMS Relay Using Tungsten VIA Layer. *Micromachines* **2018**, *9*, 579. [CrossRef] [PubMed]
9. Jo, H.C.; Choi, W.Y. Encapsulation of NEM Memory Switches for Monolithic-Three-Dimensional (M3D) CMOS–NEM Hybrid Circuits. *Micromachines* **2018**, *9*, 317. [CrossRef] [PubMed]
10. Li, X.; Hu, J.; Liu, X. A High-Performance Digital Interface Circuit for a High-Q Micro-Electromechanical System Accelerometer. *Micromachines* **2018**, *9*, 675. [CrossRef] [PubMed]
11. Islam, M.S.; Wei, R.; Lee, J.; Xie, Y.; Mandal, S.; Feng, P.-L. A Temperature-Compensated Single-Crystal Silicon-on-Insulator (SOI) MEMS Oscillator with a CMOS Amplifier Chip. *Micromachines* **2018**, *9*, 559. [CrossRef] [PubMed]

micromachines

MDPI

Article

Towards an Ultra-Sensitive Temperature Sensor for Uncooled Infrared Sensing in CMOS–MEMS Technology

Hasan Göktaş

Electrical and Electronic Engineering, Harran University, Şanlıurfa 63000, Turkey; hgoktas.gwu@gmail.com; Tel.: +90-414-318-3000

Received: 10 January 2019; Accepted: 1 February 2019; Published: 6 February 2019

Abstract: Microbolometers and photon detectors are two main technologies to address the needs in Infrared Sensing applications. While the microbolometers in both complementary metal-oxide semiconductor (CMOS) and Micro-Electro-Mechanical Systems (MEMS) technology offer many advantages over photon detectors, they still suffer from nonlinearity and relatively low temperature sensitivity. This paper not only offers a reliable solution to solve the nonlinearity problem but also demonstrate a noticeable potential to build ultra-sensitive CMOS–MEMS temperature sensor for infrared (IR) sensing applications. The possibility of a $31\times$ improvement in the total absolute frequency shift with respect to ambient temperature change is verified via both COMSOL (multiphysics solver) and theory. Nonlinearity problem is resolved by an operating temperature sensor around the beam bending point. The effect of both pull-in force and dimensional change is analyzed in depth, and a drastic increase in performance is achieved when the applied pull-in force between adjacent beams is kept as small as possible. The optimum structure is derived with a length of 57 μm and a thickness of 1 μm while avoiding critical temperature and, consequently, device failure. Moreover, a good match between theory and COMSOL is demonstrated, and this can be used as a guidance to build state-of-the-art designs.

Keywords: CMOS; MEMS; microresonators; microelectromechanical systems; thermal detector; temperature sensor; infrared sensor; microbolometer

---

1. Introduction

Microbolometers offer many advantages with their compact size, low power, capability of working at room temperature, small cost, reliable and simpler fabrication technique over bulky or relatively expensive detectors (liquid-nitrogen cooled HgCdTe (MCT), [1] etc.) in Infrared (IR) Sensing application. Ideal microbolometers should consist of high sensitivity temperature sensors and an IR absorbing layer. The IR absorbing layer converts the incident radiation into heat, and that heat is converted into the electrical signal via a temperature sensor (non-resonant [2,3], resonant-sensing [4–9]). The resonant-sensing type sensor has many advantages over the non-resonant type, such as smaller dimension and relatively low noise, due to a high-quality factor of 2.4×10^6 [9] and 1 million [10]. That is why resonant-sensing type sensors are also popular in mass sensing [11–13], but are mostly fabricated in Micro-Electro-Mechanical Systems (MEMS) technology (MEMS resonators) [5–9] rather than in complementary metal-oxide semiconductor (CMOS) technology (CMOS-MEMS resonators [4,14]).

A high-density focal plane array (FPAs) are very demanding for high-quality thermal imaging, and this requires a high-density integrated circuit (IC). It can be achieved by either building thermal detectors and IC on the same chip (CMOS–MEMS) [15,16] or bonding a separate IC and MEMS chip together [17]; however, the one that requires bonding brings extra fabrication costs and complexity. That is why CMOS–MEMS resonant-sensing type uncooled IR detectors are becoming more attractive, as they offer all-in-one (IC + MEMS), cost-effective and high sensitivity solution together. The main performance parameter for resonant-sensing type temperature sensors (cantilever, tuning fork, free–free beam, and fixed–fixed beam) is the temperature coefficient of frequency (TCF) that represents the magnitude of frequency shift (FS) with respect to the temperature change. The wide range frequency tuning capability of a fixed–fixed beam in comparison to other resonant-sensing types was demonstrated for the first time in [14] and later used in [4] to build a high sensitivity temperature sensor in CMOS technology. Moreover, fixed–fixed beam type CMOS–MEMS resonator [4] has the potential to offer high performance with their relatively high TCF (4537 ppm/K (Table 1)), while enabling a more reliable and simpler fabrication process. Despite their relatively large TCF, fixed–fixed beam type CMOS–MEMS resonators suffer from a nonlinearity problem and still need to have larger TCF for ultra-sensitive uncooled IR detection application.

In this work, the nonlinearity problem of fixed–fixed type CMOS–MEMS resonator is resolved by operating the resonator around the beam bending point. In addition, at least $31 \times$ (343 kHz/11 kHz) improvement in total absolute FS with an absolute $|TCF| > 589{,}698$ ppm/K are achieved according to COMSOL and theory for 57 µm long CMOS–MEMS resonator. $|TCF|$ increases from 589,698 ppm/K to 2178,946 ppm/K when applied Joule-heating (Vth) changes from 3.3252 V to 3.3476 V according to COMSOL. Here both Joule-heating and the change in the ambient temperature are applied together in contrast to [4], where only the ambient temperature change was used to derive $|TCF|$. Moreover, the effect of the pull-in force between two adjacent beams is studied in detail to find the optimum resonator working parameters for the sake of larger $|TCF|$. The $|TCF|$ drastically decreases from 2,333,771 ppm/K to 16,185 ppm/K when pull-in force increases from 7 MPa to 10,000 MPa according to COMSOL for 120 µm long CMOS–MEMS resonator due to decreases in thermal stress on both fixed ends. In addition, in contrast to [4], there is no thickness effect on FS while a shorter beam results in larger FS where the beam just starts to bend. The maximum temperature around beam bending point for 57 µm long beam is calculated as 530 K via COMSOL, and that does not exceed the maximum allowable temperature in CMOS–MEMS technology [18]. According to COMSOL and theory, a significant improvement in $|TCF|$ for 57 µm long CMOS–MEMS resonator over previous works can be achieved (Table 1)

Table 1. Performance comparison between this work and literature. TCF–temperature coefficient of frequency, CMOS–complementary metal-oxide semiconductor, MEMS–Micro-Electro-Mechanical Systems, NEMS–Nano Electromechanical Systems.

Design	Resonance Frequency	Absolute |TCF| (ppm/K)	Technology
This work (57 µm long CMOS–MEMS Resonator)	1.92 MHz	2,178,946	CMOS–MEMS
120 µm long CMOS–MEMS Resonator [4]	640 kHz	4537	CMOS–MEMS
AlN Piezoelectric Nanomechanical Resonator [5]	161.4 MHz	30	NEMS
Nanomechanical Torsional Resonator [6]	842 kHz	548	NEMS
Silicon Micromechanical Resonator [7]	101 MHz	29.7	MEMS

2. Fabrication

The CMOS–MEMS resonators can be fabricated via a post-process followed after a CMOS 0.6 µm process that includes a CHF_3/O_2 process for SiO_2 etching between adjacent beams and XeF_2 process for Silicon etching underneath the beams [14].

In this study, the device structures (Figure 1) are slightly changed for the sake of better performance. However, the distance between devices and the silicon etching ratio is kept the same.

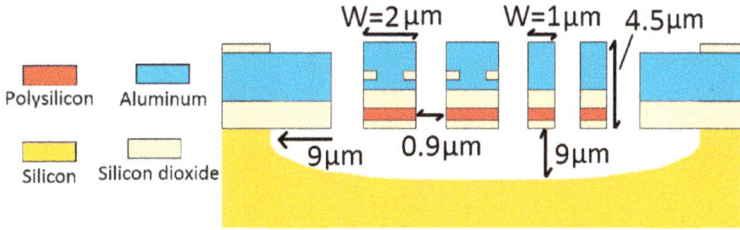

Figure 1. The cross section for Device 1 (W = 2 μm) and for Device 2 (W = 1 μm), where W is the thickness.

3. Theory Modeling and Optimization

The working principle of the CMOS–MEMS resonator (Figure 1) is based on pull-in force (via DC bending voltage (Vdc) applied between two adjacent beams), and the Joule-heating voltage (Vth) applied on the embedded heater (polysilicon layer) through the resonant beam. Pull-in force enables the softening effect on the resonant beam and, consequently, starts the resonance operation while Joule-heating increases the temperature throughout the resonant beam and resultes in relatively high thermal stress on the fixed ends. This Joule-heating effect causes a wide range of frequency tuning and this was first time demonstrated in [14]. The resonance frequency with respect to axial load [19] is:

$$f = \frac{4.73^2}{2\pi L^2} \left(1 + \frac{PL^2}{EI\pi^2}\right)^{\frac{1}{2}} \left(\frac{EI}{m}\right)^{\frac{1}{2}} (1) \tag{1}$$

where I is the moment of inertia, L (m) is the length, m (kg/m) is the mass per unit length, and P is the total compressive axial load on fixed ends [20]. More detail is given in [18]. In addition to Equation (1), COMSOL was used to build the CMOS–MEMS resonators (Figure 1) and calculate their resonance frequency responses with respect to temperature. The simulation environment was selected as a vacuum, and ambient temperature (Tamb) was set to 273 K. Solid mechanics, heat transfer, and electric currents tools were combined together in multiphysics to couple heat transfer with solid mechanics and electric currents. Mesh study was conducted to find the optimum mesh set up for the simulation. Both the "extremely fine mesh" and "fine mesh" were compared to decrease time budget, where tetrahedral meshing was used throughout the structure. There was only a slight change observed between the results. Polysilicon conductance was set as 1.16×10^5 S/m as it was already measured and verified [18]. Electric current was used to heat the beams via Joule-heating while the heat transfer module was used to model temperature distribution throughout the beam and solid mechanics was used to model deformation and mode shapes.

The resonance frequency tuning range with the application of Joule-heating was around 761 kHz when the pull-in force was 7 MPa, and it was around 276.5 kHz when it was 10,000 MPa (Figure 2a). This is attributed to the fact that both the pull-in force and Joule-heating results in beam bending. Pull-in force, however, created an ignorable stress on the fixed ends in comparison to Joule-heating and consequently results in a very small frequency tuning range [21]. In another words, the bending should be resulted mainly because of thermal stresses (Joule-heating) while keeping the pull-in force as minimum as possible to get the maximum frequency tuning range.

The slope of the resonance frequency with respect to the applied Joule-heating voltage (Vth) was not constant but kept on increasing (α4 > α3 > α2 > α1) (Figure 2a) with an increase in temperature. This nonlinear effect was first observed in [14], and allows better FS at higher temperatures (Figure 2b) and consequently enables higher sensitivity temperature sensor design. This effect was analyzed partially in [4], and the temperature sensitivity was found as 2.98 kHz/C without any Joule-heating application.

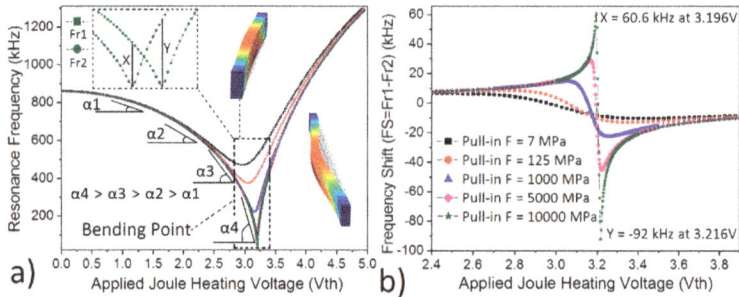

Figure 2. The effect of pull-in force (F) on the (**a**) Frequency tuning and (**b**) frequency shift (FS) in COMSOL simulation for Device 1 for a length of 120 µm long fixed–fixed beam, where Fr1 and Fr2 are the resonance frequency responses with ambient temperature of Tamb and Tamb + 1 K respectively.

In contrast to [4,14], here we studied the FS in detail by combining both ambient temperature (Tamb) change and Joule-heating for highly sensitivity temperature sensors in microbolometer application. This required the full analysis of the frequency response (Figure 2a) where the resonance frequency decreases until it reaches the bending point and then starts to increase. Two different resonance frequency (Fr1, Fr2) responses with respect to applied Joule-heating voltage were calculated via COMSOL at two different environment temperature (Tamb1 = 273 K and Tamb2 = 274 K) in Figures 2–4. Hence, FS for 1 Kelvin change can be derived by subtracting resonance frequency responses (Fs = Fr1−Fr2) for every applied Vth (Figures 2b, 3 and 4). The optimum device operation point (larger FS, consequently better sensitivity) was found around the bending point (Figure 2a) where the beam was just starting to have a 0.38 µm bending and gives maximum FS values (X and Y). The FS was 5 kHz when Vth = 0 V and reaches up to 60.6 kHz when Vth = 3.196 V and −92 kHz when Vth was 3.216 V. If Vth is switched from 3.196 V and 3.216 V, then total absolute FS will be X+|Y| = 152.6 kHz (Figure 2b). Moreover, the pull-in force should be as small as possible to create as sharp a bending curve as possible (Figure 2a). This, in turn, would result in a larger X and |Y| value and consequently larger FS and much better sensitivity. The total absolute FS was around 152.6 kHz (|TCF| = 2,333,771 ppm/K at Vth = 3.216 V) when pull-in force was 7 MPa whereas it was around 16.8 kHz (|TCF| = 16,185 ppm/K at Vth = 3.425 V) when pull-in force was 10,000 MPa.

Further optimization was conducted by analyzing the dimensional effect to find the optimum structure for the sake of larger FS. The Joule-heating is studied in Figure 3b for Device 1 to study the effect of thickness on FS and in Figure 4b for Device 2 to study the effect of length on FS by using COMSOL. In the same way, uniform heating was applied to the beams to calculate the FS via Equation (1) in Figures 3a and 4a. That is why max temperature is used to plot FS in Figures 3b and 4b in contrast to the uniform temperature profile in Figures 3a and 4a. The minimum pull-in force was applied to every beam in Figures 3 and 4 to get the largest FS. A good match between COMSOL and Equation (1) is achieved for the total absolute FS.

The CMOS–MEMS resonator's width can go up to 6 µm with a metal-3 layer and can go up to 5.1 µm [14] after post-processing. That is why the thickness should not exceed 4 µm. Otherwise, devices cannot resonate. In this study, we set the width as 4.5 µm (Figure 1) and, hence, only three different thickness profiles were used. The thinner the beam, the larger the FS at relatively low temperature (T < 285 K) as was demonstrated in [4]. However, this behavior changed with the increase in temperature (Figure 3). The thickness has almost no effect on the FS at the bending point according to Equation (1) and COMSOL. The total absolute FS was 146.5 kHz when the beam thickness is 1 µm, and it was around 142.7 kHz when the thickness was 3 µm according to COMSOL. In the same way, it is 168.4 kHz when the thickness is 1 µm, and 164.8 mkHz when the thickness is 3 µm according to (1). Although there was no noticeable change in the FS, the thinner beam was preferable due to requiring less temperature for bending ($T_{\text{bending point}}$ = 312 K, Figure 3b) and, consequently, has smaller thermal stresses [18]. That is why the length study is conducted for 1 μm thick beam (Device 2) in Figure 4.

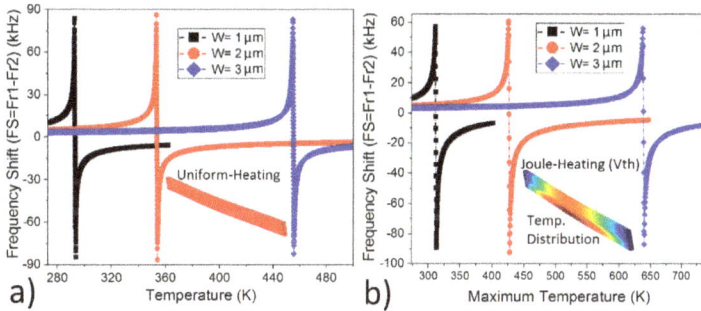

Figure 3. Frequency Shift (FS) with respect to 1 Kelvin (K) change by (**a**) Equation (1), and (**b**) COMSOL, when thickness (W) changes from 1 μm to 3 μm for Device 1 with a device length of 120 μm.

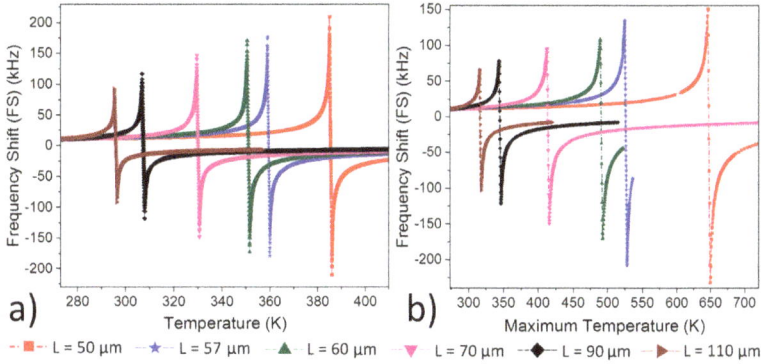

Figure 4. Frequency Shift (FS) with respect to 1 Kelvin (K) change by (**a**) Equation (1), and (**b**) COMSOL, when length (L) changes from 50 μm to 110 μm for Device 2 with a device thickness of 1 μm.

The minimum length was set as 50 μm and the maximum one was set as 110 μm for the following reasons; 50 μm beam already exceeded the temperature limit (Figure 4b, $T_{\text{bending point}}$ = 650 K) that the CMOS layers could tolerate and 110 μm beam was at the limit of stiction risk in post fab process due to sacrificing low stiffness constant. FS increased with the increase in length at relatively low temperatures (Figure 4), and this is attributed to the fact that the longer beam has higher TCF values [4,14]; however, this is only valid before the bending point. Once the beam reaches the bending point, the shorter beam results in larger FS and consequently better sensitivity. The total absolute FS increased from 169.9 kHz to 382 kHz according to COMSOL and it is increased from 184.4 kHz to 419 kHz according to (1) when the length decreased from L = 110 μm to L = 50 μm

There is no study conducted on the effect of width on temperature sensitivity because CMOS is not a custom process and the number of layers and their thicknesses are well defined. In addition, highly sensitivity temperature sensors require material with high thermal expansion constant, such as aluminum layers, and this eliminates the possibility of changing the width.

The optimum structure is shaped according to the results obtained from Figures 3 and 4 with a length of 57 μm and a thickness of 1 μm (Device 2). The total absolute FS is 343 kHz (|TCF| = 589,698 ppm/K at Vth = 3.3252 V, |TCF| = 2,178,946 ppm/K at Vth = 3.3476 V) where the maximum temperature around bending point is 530 K with a 0.14 μm bending. The optimum structure's working temperature is limited to 530 K in this work because the maximum allowable

Micromachines **2019**, *10*, 108

temperature for the similar structure in CMOS process was found to be around 530 K when 5.7 V and 17.4 mW was applied on embedded polysilicon layer [18]. The final structure's mesh was set to "extremely fine mesh" with a very high-density sweep of Vth (0.0004 V resolution) to get the maximum accuracy in the results. The good match is achieved between COMSOL and (1); total absolute FS is 343 kHz according to COMSOL, and it is 356 kHz according to (1).

The 0.14 μm thermal bending offers the potential for a high-density thermal detector array in CMOS. The total improvement of resonator's sensitivity with respect to temperature can be derived from the ratio of the total absolute FS with Joule-heating application (X + |Y| (Figure 2b)) over the FS without any Joule-heating application (at Vth = 0 V). FS at Vth = 0 V is 11.2 kHz, and total absolute FS is 343 kHz (Figure 4b) for 57 μm long beam, and this brings around a 31× improvement in the overall sensitivity.

4. Conclusions

Fixed–Fixed beam type CMOS–MEMS resonator was studied in detail and optimized to build the state-of-the-art temperature sensors for high-performance uncooled microbolometers. The best performance was achieved with 57 μm long and 1 μm thick fixed–fixed beam with a maximum temperature of around 530 K, that is close but still under the critical temperature in CMOS technology [18]. The total frequency shift increased from 11 kHz to 343 kHz (31×) for 57 μm beam with much larger |TCF| (2,178,946 ppm/K) while keeping the pull-in force application as small as possible. Furthermore, the nonlinearity problem of fixed–fixed beam type CMOS–MEMS resonator was addressed by operating the device around the beam bending point. A good match between COMSOL and theory was demonstrated and can be used as guidance in future researches to build an ultra-sensitive temperature sensor for microbolometers in CMOS technology. This in return, can enable a less expensive, compact, and wider range of application compatibility such as internet of things.

Funding: This research was funded by Scientific Research Project Foundation of Turkey (grant number 18073).

Acknowledgments: The author especially wishes to thank COMSOL for their support in setting up the simulation environment accurately for CMOS–MEMS resonator in this study.

Conflicts of Interest: The authors declare no conflict of interest.

References

1. Marsili, F.; Verma, V.B.; Stern, J.A.; Harrington, S.; Lita, A.E.; Gerrits, T.; Vayshenker, I.; Baek, B.; Shaw, M.D.; Mirin, R.P.; et al. Detecting single infrared photons with 93% system efficiency. *Nat. Photonics* **2013**, *7*, 210–214. [CrossRef]
2. Chen, C.; Yi, X.; Zhao, X.; Xiong, B. Characterization of VO$_2$ based uncooled microbolometer linear array. *Sens. Actuators A Phys.* **2001**, *90*, 212–214. [CrossRef]
3. Kang, D.H.; Kim, K.W.; Lee, S.Y.; Kim, Y.H.; Keun Gil, S. Influencing factors on the pyroelectric properties of Pb (Zr, Ti) O$_3$ thin film for uncooled infrared detector. *Mater. Chem. Phys.* **2005**, *90*, 411–416. [CrossRef]
4. Göktaş, H.; Turner, K.L.; Zaghloul, M.E. Enhancement in CMOS-MEMS Resonator for High Sensitive Temperature Sensing. *IEEE Sens. J.* **2017**, *17*, 598–603. [CrossRef]
5. Hui, Y.; Gomez-Diaz, J.S.; Qian, Z.; Alù, A.; Rinaldi, M. Plasmonic piezoelectric nanomechanical resonator for spectrally selective infrared sensing. *Nat. Commun.* **2016**, *7*, 11249. [CrossRef] [PubMed]
6. Zhang, X.C.; Myers, E.B.; Sader, J.E.; Roukes, M.L. Nanomechanical Torsional Resonators for Frequency-Shift Infrared Thermal Sensing. *ACS Nano Lett.* **2013**, *13*, 1528–1534. [CrossRef] [PubMed]
7. Gokhale, V.J.; Rais-Zadeh, M. Uncooled Infrared Detectors Using Gallium Nitride on Silicon Micromechanical Resonators. *IEEE Micromech. Syst.* **2014**, *23*, 803–810. [CrossRef]
8. Hui, Y.; Rinaldi, M. Fast and high-resolution thermal detector based on an aluminum nitride piezoelectric microelectromechanical resonator with an integrated suspended heat absorbing element. *Appl. Phys. Lett.* **2013**, *102*, 093501. [CrossRef]
9. Larsen, T.; Schmid, S.; Grönberg, L.; Niskanen, A.O.; Hassel, J.; Dohn, S.; Boisen, A. Ultrasensitive string-based temperature sensors. *Appl. Phys. Lett.* **2011**, *98*, 121901. [CrossRef]

10. Tao, Y.; Boss, J.M.; Moores, B.A.; Degen, C.L. Single-crystal diamond nanomechanical resonators with quality factors exceeding one million. *Nat. Commun.* **2014**, *5*, 3638. [CrossRef] [PubMed]

11. Jensen, K.; Kim, K.; Zettl, A. An atomic-resolution nanomechanical mass sensor. *Nat. Nanotechnol.* **2008**, *3*, 533–537. [CrossRef] [PubMed]

12. Yang, Y.T.; Callegari, C.; Feng, X.L.; Ekinci, K.L.; Roukes, M.L. Zeptogram-Scale Nanomechanical Mass Sensing. *ACS Nano Lett.* **2006**, *6*, 583–586. [CrossRef] [PubMed]

13. Baek, I.B.; Byun, S.; Lee, B.K.; Ryu, J.H.; Kim, Y.; Yoon, Y.S.; Jang, W.I.; Lee, S.; Yu, H.Y. Attogram mass sensing based on silicon microbeam resonators. *Nat. Sci. Rep.* **2017**, *7*, 46660. [CrossRef] [PubMed]

14. Göktaş, H.; Zaghloul, M.E. Tuning In-Plane Fixed–Fixed Beam Resonators with Embedded Heater in CMOS Technology. *IEEE Electron Dev. Lett.* **2015**, *36*, 189–191. [CrossRef]

15. Escorcia, I.; Grant, J.P.; Gough, J.; Cumming, D. Terahertz Metamaterial Absorbers Implemented in CMOS Technology for Imaging Applications: Scaling to Large Format Focal Plane Arrays. *IEEE J. Sel. Top. Quantum Electron.* **2017**, *23*, 4700508. [CrossRef]

16. Eminoglu, S.; Tanrikulu, M.Y.; Akin, T. A Low-Cost 128 × 128 Uncooled Infrared Detector Array in CMOS Process. *IEEE J. Microelectromech. Syst.* **2008**, *17*, 20–30. [CrossRef]

17. Forsberg, F. CMOS-Integrated Si/SiGe Quantum-Well Infrared Microbolometer Focal Plane Arrays Manufactured with Very Large-Scale Heterogeneous 3-D Integration. *IEEE J. Sel. Top. Quantum Electron.* **2015**, *21*, 2700111. [CrossRef]

18. Göktaş, H.; Zaghloul, M.E. The implementation of low-power and wide tuning range MEMS filters for communication applications. *Radio Sci.* **2016**, *51*, 1636–1644. [CrossRef]

19. Jha, C.M. Thermal and Mechanical Isolation of Ovenized MEMS Resonator. Ph.D. Thesis, Department of Mechanical Engineering, Stanford University, Palo Alto, CA, USA, 2008.

20. Abawi, A.T. The Bending of Bonded Layers Due to Thermal Stress. Available online: http://hlsresearch.com/personnel/abawi/papers/bend.pdf (accessed on 23 October 2014).

21. Hopcroft, M.A. Temperature-Stabilized Silicon Resonators for Frequency References. Ph.D. Thesis, Department of Mechanical Engineering, Stanford University, Palo Alto, CA, USA, 2007.

micromachines

MDPI

Article

A High-Performance Digital Interface Circuit for a High-Q Micro-Electromechanical System Accelerometer

Xiangyu Li [1], Jianping Hu [1,*] and Xiaowei Liu [2]

[1]   Faculty of Information Science and Technology, Ningbo University, Ningbo 315211, China; lixiangyu@nbu.edu.cn
[2]   MEMS Center, Harbin Institute of Technology, Harbin 150001, China; liuxiaowei3@outlook.com
*   Correspondence: hujianping2@nbu.edu.cn; Tel.: +86-0574-87600346

Received: 11 November 2018; Accepted: 18 December 2018; Published: 19 December 2018

Abstract: Micro-electromechanical system (MEMS) accelerometers are widely used in the inertial navigation and nanosatellites field. A high-performance digital interface circuit for a high-Q MEMS micro-accelerometer is presented in this work. The mechanical noise of the MEMS accelerometer is decreased by the application of a vacuum-packaged sensitive element. The quantization noise in the baseband of the interface circuit is greatly suppressed by a 4th-order loop shaping. The digital output is attained by the interface circuit based on a low-noise front-end charge-amplifier and a 4th-order Sigma-Delta ($\Sigma\Delta$) modulator. The stability of high-order $\Sigma\Delta$ was studied by the root locus method. The gain of the integrators was reduced by using the proportional scaling technique. The low-noise front-end detection circuit was proposed with the correlated double sampling (CDS) technique to eliminate the $1/f$ noise and offset. The digital interface circuit was implemented by 0.35 μm complementary metal-oxide-semiconductor (CMOS) technology. The high-performance digital accelerometer system was implemented by double chip integration and the active interface circuit area was about 3.3 mm × 3.5 mm. The high-Q MEMS accelerometer system consumed 10 mW from a single 5 V supply at a sampling frequency of 250 kHz. The micro-accelerometer system could achieve a third harmonic distortion of −98 dB and an average noise floor in low-frequency range of less than −140 dBV; a resolution of 0.48 μg/Hz$^{1/2}$ (@300 Hz); a bias stability of 18 μg by the Allen variance program in MATLAB.

Keywords: MEMS; interface circuit; high-Q capacitive accelerometer; Sigma-Delta

1. Introduction

Capacitive accelerometers are widely used in the military and civilian fields because of their low power consumption, simple structure, good stability and easy integration with the complementary metal-oxide-semiconductor (CMOS) process [1]. In recent years, high-performance capacitive accelerometers with an accuracy of sub-μg level occupy a large market share in inertial navigation, space microgravity measurement, platform stability control and other fields. The micro-accelerometers with an open-loop output have a simple structure, but the signal bandwidth is limited by the sensitive structure and the input range of the signal is greatly reduced [2–4]. Therefore, the micro-accelerometers usually work in a closed-loop feedback state to obtain better linearity, dynamic range and signal bandwidth. The closed-loop working mode can also increase the electrical damping of the mechanical structure and improve effectively its electrical response [5,6]. High over sampling rate (OSR), high-order topology and multi-bit quantization are used to improve the noise shaping ability of Sigma-Delta ($\Sigma\Delta$) micro-accelerometers.

A large OSR requires high sampling frequency, which leads to coupling between different noise sources and increasing power consumption in $\Sigma\Delta$ micro-accelerometers. The high-Q sensitive structure introduces a large phase shift at the resonance frequency, and the stability of the whole high order system will be greatly reduced. If a high-Q sensitive structure is used to reduce mechanical noise and a high-order structure is used to reduce quantization noise, the problem of system stability will become a major problem. It is necessary that cascading a phase compensator after the front-end charge amplifier to provide additional phase compensation, which is equivalent to providing electrical damping to the under-damped mechanical structure to stabilize the loop. In other literature, phase compensators are placed in the feedback loop, which can improve the feedforward path gain, but this can also reduce the gain in the feedback path and reduce the input dynamic range. It is difficult to design a linear micro-accelerometer with a multi-bit quantization structure because the signal conversion process of the sensitive structure is nonlinear. At present, the main research on the interface circuit of micro-accelerometers is still based on a low-Q sensitive structure, low-order $\Sigma\Delta$ system and a one-bit feedback structure [7,8]. The micro-accelerometers with analog output can achieve a high precision output of less than 1 $\mu g/Hz^{1/2}$, but the performance of digital closed-loop micro-accelerometers reported is difficult to achieve a precision at the sub-μg level [9]. The digital micro-accelerometers with sub-μg precision output has a lot of application requirements in the field of geophone, national defense and military. Therefore, the noise theory, system stability analysis and key technology of high-precision closed-loop micro-accelerometers are mainly studied in this paper, which is aimed at realizing a high-performance interface circuit chip with sub-μg accuracy.

The high-Q accelerometer sensitive element, front-end charge sensing circuit, sample and hold circuit, phase compensation circuit and high-order Sigma-Delta modulator circuit are introduced and designed in Section 2. In Section 3, we show a detailed analysis based on the noise characteristics and stability of micro-accelerometers with an application specific integrated circuit (ASIC) interface. The performance can be improved by a correlated double sampling (CDS) technique and a proportional scaling technique. The performance parameters of micro-accelerometers were tested by the experiments. Finally, Section 4 concludes the study of a high-Q MEMS accelerometer with a high-precision integrated circuit and testing results, which show that the performance level of micro-accelerometers in this work has great advantages in the application of inertial navigation and the nano-satellites field by comparison.

2. Materials and Methods

2.1. Materials

The high-Q sensitive structure which is encapsulated in vacuum is from Colibrys Company (Neuchatel, Switzerland). The interface circuit based on micro-accelerometers was fabricated by a 0.35 μm CMOS process and cooperated with Shanghai Huahong Integrated Circuit (Shanghai, China).

2.2. High-Q Accelerometer Sensitive Element

The equivalent bridge model of the vacuum packaged silicon micro-accelerometers is shown in Figure 1. The upper and lower capacitance plates in Figure 1 are fixed plates and the equivalent variable capacitors C_{S1} and C_{S2} are formed between the mass and the plates. C_{P1} and C_{P2} are parasitic capacitors. When the external acceleration acts on the sensitive element, the displacement of the mass will change, which is relative to the plates. This can result in the corresponding change of the variable capacitance. The change of the two equivalent sensitive capacitances will be perceived by the post-detection circuit. The accelerometer sensitive element with vacuum packaged silicon structure used for design, simulation and test in this paper was obtained from Colibrys Company (SF1500). The sensitive element could achieve an open-loop resonant frequency of 1 kHz, a high-quality factor of more than 30 and a Brownian noise corresponding of an equivalent acceleration of less than

60 ng/Hz$^{1/2}$. The corresponding static capacitance and the sensitivity of the sensor element were 180 pF and 10 pF/g. Major parameter indicators are shown as in Table 1.

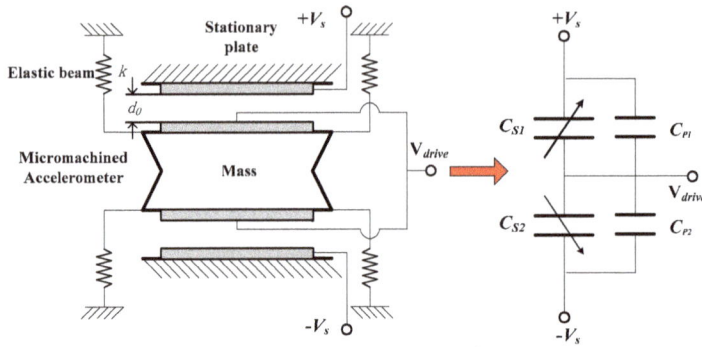

Figure 1. Vacuum-packaged bulk micro-accelerometer and equivalent bridge model.

Table 1. Parameters of the high-Q sensor.

Parameters	Value
Sensitivity	10 pF/g
Proof Mass (m)	6.2×10^{-7} kg
Rest Capacitance (C_0)	180 pF
Damping Coefficient (b)	0.01 N/(m·s)
Sensing Gap Distance (d)	2 μm
Resonance Frequency (ω_0)	1000 Hz
Quality Factor (Q)	>30
Brownian Noise Floor	<60 ng/Hz$^{1/2}$

Figure 2 shows the differential capacitance model of the sensitive structure, in which d was the distance between the upper and lower plates. When the mass is in equilibrium and the two differential capacitance values are equal, the static capacitance is shown as follows:

$$C_0 = \frac{\varepsilon \varepsilon_0 A}{d} \tag{1}$$

ε_0—the vacuum dielectric constant
ε—the relative dielectric constant between the sensitive capacitor plates
A—the positive area of the sensitive capacitor plates
x—the displacement of the sensitive mass block under the external acceleration
When the displacement of the sensitive mass causes changes in differential capacitance pairs, the variable capacitance C_{S1} and C_{S2} in Figure 2 can be expressed respectively:

$$C_{S1} = \frac{\varepsilon \varepsilon_0 A}{d - x} = \frac{C_0}{1 - \frac{x}{d}} \tag{2}$$

$$C_{S2} = \frac{\varepsilon \varepsilon_0 A}{d + x} = \frac{C_0}{1 + \frac{x}{d}} \tag{3}$$

In the closed-loop system, the displacement of the sensitive mass was very small relative to the plate spacing. The relative variation of the capacitance (ΔC) can be written as follows:

$$\Delta C = C_{S1} - C_{S2} = \frac{\varepsilon \varepsilon_0 A}{d - x} - \frac{\varepsilon \varepsilon_0 A}{d + x} \approx 2C_0 \frac{x}{d} \tag{4}$$

It can be seen that the relative displacement of the mass and the input acceleration signal are approximately linear in the input signal band, which was much smaller than the resonant frequency of the mechanical structure. That is $x \approx \frac{a}{\omega_0^2}$, where a denotes the acceleration signal and ω_0 denotes the mechanical resonance frequency. The Equation (4) can be expressed as:

$$a = \frac{\Delta C d\omega_0^2}{2C_0} \qquad (5)$$

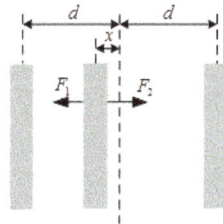

Figure 2. Differential capacitance model of the sensitive element.

2.3. High-Order Interface Circuit Based on Micro-Accelerometers

The closed-loop micro-accelerometers use the feedback principle of electrostatic force to confine the sensitive mass to the balance position, which greatly reduces the sensitive mass's displacement in order to reduce the nonlinear error in charge conversion and improve the overall linearity, bandwidth and amplitude range of the input acceleration signal. In this paper we propose a high-precision digital micro-accelerometer with sub-µg noise level with a high-Q sensitive structure encapsulated in a vacuum, which was used to reduce the mechanical noise. The high-order noise shaping ability was realized by combining the high-order topological structure. Due to underdamping, slow corresponding output in response and poor seismic performance of high vacuum mechanical structures, there will be stability problems after constituting a high-order system with the ΣΔ modulator. Therefore, when the system of micro-accelerometers can achieve sub-µg noise level, the stability of the system should be fully considered in the digital interface circuit design of micro-accelerometers [10–13]. Aiming at the stability problem of high-precision digital micro-accelerometer interface circuit, a phase compensator circuit can be designed to provide phase compensation, enhance electrical damping and improve the system response. In addition, in order to overcome the influence of process error on the stability of high-order interface circuit, reasonable circuit design and parameter optimization are needed.

Figure 3 shows a diagram of the front-stage charge-sensitive circuit. In this paper, we propose a fully differential switched-capacitor detection circuit, in which C_R is the reference capacitor and C_f is the integral capacitor. The front-stage sensing circuit consists of an equivalent mechanical structure, a reference capacitor pair, a charge-sensitive and a correlated double-sampling and holding module. The output voltage of the charge sensitive circuit can be expressed as:

$$V_{out} = \frac{2V_r \Delta C}{C_f} = \frac{4C_0 V_r}{C_f d\omega_0^2} a(f) \qquad (6)$$

The input acceleration signal is converted into the voltage signal of the front-stage sensitive circuit. In Equation (6), V_r is the reference voltage. The sensitivity of the detection was limited by the initial capacitance of the sensitive structure, the distance between the plates and the resonant frequency of the mechanical structure. In this paper the static capacitance value of the sensitive structure and the reference capacitance were 180 pF respectively. An additional capacitor can be connected in parallel with the sensitive structure to increase the equivalent static capacitance value. But the static capacitance can't be increased indefinitely, which will affect the loop stability and the accuracy of charge conversion. The timing diagram of the front-end circuit is as shown in Figure 3b. There are five phases in operation of the circuit, which is the reset phase, charge sensing phase A, charge sensing phase B, sampling phase and electrostatic force feedback phase. The switch S4_inv and S5_inv were the reverse clock of S4 and S5, respectively. Electrostatic force feedback and charge sensitivity operate at different times of a cycle to eliminate noise coupling between them. In the reset phase, the input electrode voltage of the interface was reset to ensure a correct bias point and the capacitor was discharged to erase the memory from the previous cycle. A small size of switch S6 was designed to reduce charge injection. In the charge sensing phase A, the reference voltages $+V_s$ and $-V_s$ were applied to the sensor mass and common electrode of the reference capacitors, respectively. The capacitor stores the amplified voltage and the error signal including the offset and noise of the operational amplifier. The output of the charge sensing is given by:

$$\Delta V_{out1} = V_{error} - V_S\frac{C_{S1} - C_{S2}}{C_f} \tag{7}$$

where C_f is the integration capacitance (10 pF). During the charge sensing phase B, the voltages of sensor mass and common electrode of the reference capacitors were kept at $+V_s$ and $-V_s$, respectively. The output of the charge sensing is expressed as:

$$\Delta V_{out2} = V_{error} + V_S\frac{C_{S1} - C_{S2}}{C_f} \tag{8}$$

The differential output of the sample and hold circuit is represented by:

$$\Delta V_{out} = \Delta V_{out2} - \Delta V_{out1} = 2V_S\frac{C_{S1} - C_{S2}}{C_f} \tag{9}$$

The values of the nominal capacitance of the sensor element and the reference capacitance were 180 pF. The integration capacitance was set to 10 pF, which was a trade-off between the noise performance and system stability. We set a pre-stage gain of 30 V/g and an accelerometer system sensitivity of 1.866 V/g. In this paper the bandwidth of the accelerometer was 300 Hz, which was defined by an increasing low-frequency noise spectral density of 3 dB.

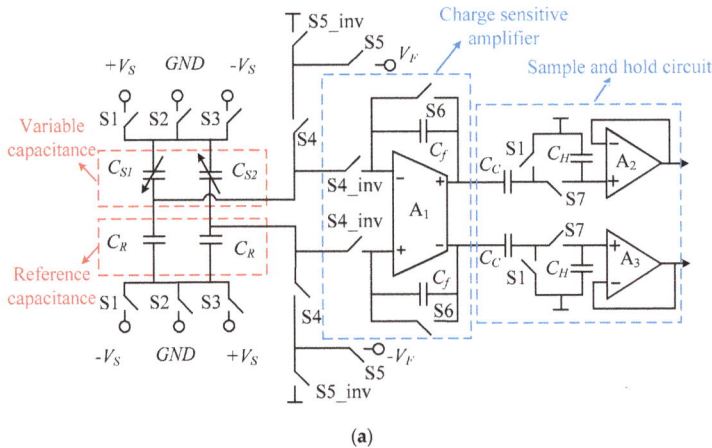

(a)

Figure 3. *Cont.*

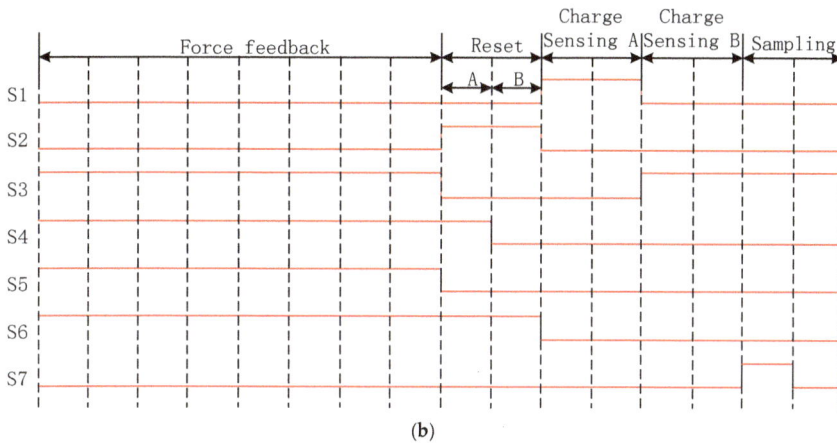

(b)

Figure 3. Front-end charge sensing circuit and timing diagram. (**a**) Front-end charge sensing circuit for micro-accelerometers; (**b**) Timing diagram for front-end charge sensing circuit.

The high-Q sensitive structure can introduce a pair of complex poles near the imaginary axis to the closed-loop filter. The high-frequency parasitic resonant modes and the complex poles can destabilize the high-Q system easily. In this paper we propose a phase compensator circuit which can introduce an extra zero to compensate for loop filters. The low-frequency loop gain control was considered based on a good noise shaping ability. In this lead compensator circuit, C_1 and C_3 had the same capacitance value. The ratio between C_2 and C_3 determined the compensation degree. For a high-Q sensitive structure, a heavy compensation was chosen. The sampling frequency of the phase compensator circuit was 250 kHz. The lead compensator with a transfer function in discrete-time z-domain can be expressed as:

$$H_{cmp}(z) = \frac{C_1}{C_3} - \frac{C_2}{C_3}z^{-1} \tag{10}$$

C_1 and C_3 have the same capacitance value and at the case of $C_2 = \alpha C_3$, the Equation (10) can be expressed as:

$$H_{cmp}(z) = 1 - \alpha z^{-1} \tag{11}$$

In Equation (11), α indicates the depth of compensation. The lead compensator operates as a proportion-derivative (PD) controller and the stability is improved by positioning the zero closer to the open-loop poles of the filter, which is resulting in an increase of the amount of phase lead. If the compensation depth is insufficient or excessive, the closed-loop system may have stability problems. For over-compensated sigma-delta accelerometer systems, the system may also be unstable if the loop gain is too small. Overcompensation of sigma-delta accelerometer systems can also affect the noise shaping ability of a post-stage modulator. Although the noise shaping ability of the modulator decreases with the increase of compensation depth, more-order structure and a high-Q sensitive structure can be used to reduce the impact of the reduction of noise shaping ability caused by depth compensation. Because of the high-order system structure in this paper, we proposed a lead compensator circuit as shown in Figure 4. The stability of the system was more important than the noise shaping ability of the modulator, so we set a depth compensation coefficient of 0.9.

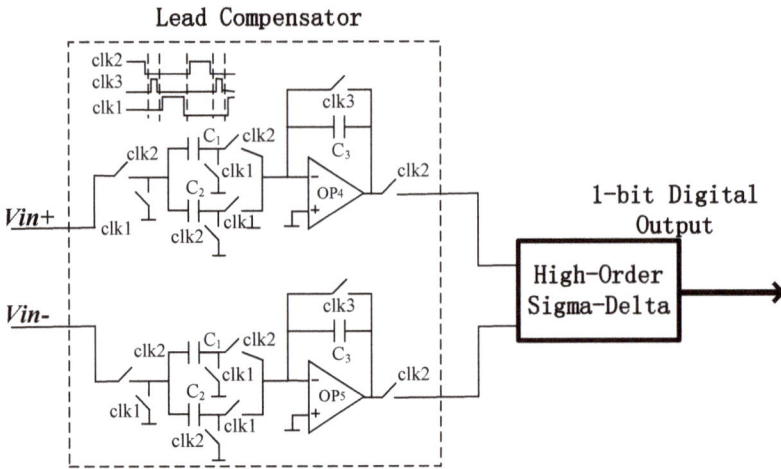

Figure 4. Lead compensator circuit.

We propose the system structure of the ΣΔ modulator as shown in Figure 5a based on stability analysis of ΣΔ micro-accelerometers. In order to achieve a better noise suppression performance at low-frequency, we used a correlated double sampling technique to improve the noise level of the first stage integrator. The one-bit quantizer was achieved by the dynamic comparator. The output of the comparator was as a control signal to control feedback reference voltage V_{ref+} and V_{ref-} in the first stage integrator [14,15]. As shown in Figure 5b, the timing diagram of the ΣΔ modulator circuit, wherein *ck*1 and *ck*2 were the two-phase non-overlapping clock, *ck*1 was active-high, *ck*2 was active-low. The shutdown time of *ck*1*d* was later than *ck*1; the shutdown time of *ck*2*d* was later than *ck*2. This could effectively suppress the influence of charge injection and clock-feedthrough in the switched-capacitor (SC) circuit. In the ΣΔ modulator circuit, the double sampling technique was also used to increase the equivalent sampling frequency in order that the sampling capacitance of the input signal and the sampling capacitance of the feedback signal were separated. The charge transfer at the integration phase is reduced and the accuracy of the integrator can be improved. In this paper we propose a topology of distributed feedback ΣΔ accelerometers with a feedforward structure. This structure combines some advantages of a feedforward and feedback topology structure and has the characteristics of good system stability and a small output signal swing. We designed the main parameters of the ΣΔ modulator as shown in Table 2.

Table 2. Parameters of the ΣΔ modulator circuit.

ΣΔ Modulator Circuit	
Loop Filter Topology	Fourth-Order Switched-Capacitor
Integration Capacitor	10 pF
Oversampling Ratio (OSR)	417
Signal-to-Noise Ratio (SNR)	108 dB
Sampling Frequency	250 kHz
Third Harmonic Distortion	−98 dB

(a)

(b)

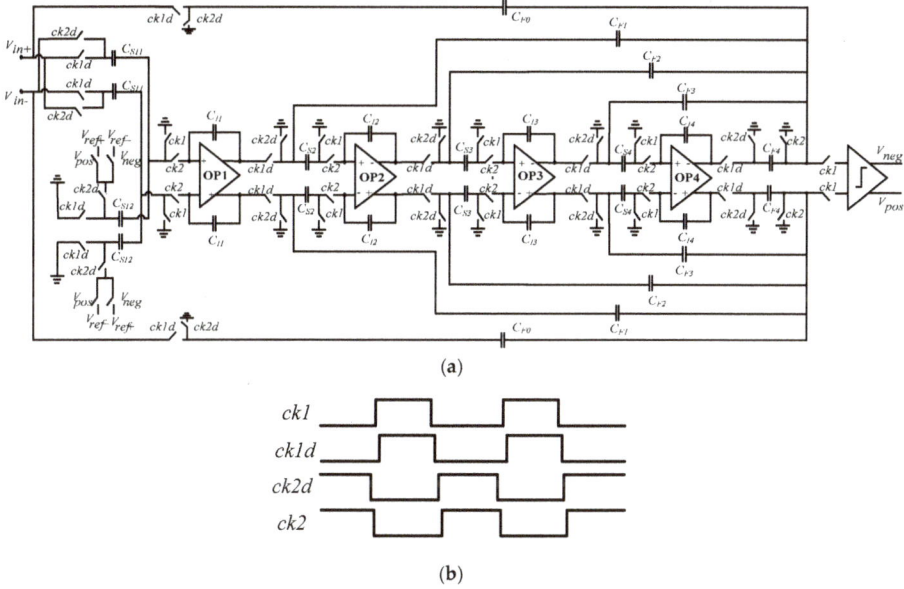

Figure 5. (**a**) High-order ΣΔ modulator circuit; (**b**) the timing diagram of ΣΔ modulator circuit.

3. Result and Discussion

3.1. Noise Characteristics and Stability Analysis of Micro-Accelerometers

In consideration of a relatively low gain at low-frequency in the feedback structure and a relatively large nonlinearity problem of the output signal. Figure 6 shows the analysis model of ΣΔ micro-accelerometers in this paper. $K_{x/V}$ in Figure 6 is the amplification factor from the displacement output of the sensitive structure to the output voltage of the charge sensitive circuit. H_c is the pre-stage phase compensator; f_{a1}, f_{a2}, f_{a3} and f_{a4} are feedforward coefficients; f_{b1}, f_{b2}, f_{b3} and f_{b4} are feedback coefficients; k_1, k_2, k_3 and k_4 are integrator gain coefficients; $K_{V/a}$ is the gain coefficient from feedback voltage to equivalent acceleration. The main noise sources introduced in the model are the Brownian noise of mechanical structure, the electrical noise of the pre-stage charge amplifier and the quantization noise of the post-stage ΣΔ. In consideration of the accuracy discreteness of the micro-accelerometer sensitive structure, there are four distributed feedback factors in the post-stage modulator circuit of the ΣΔ micro-accelerometer system in this paper. The stability of the loop can be effectively controlled by adjusting the feedback coefficient, especially adjusting the feedback coefficient f_{b1} of the first integrator. So, the local feedback factor f_{b1} is designed as an off-chip adjustable part. The low-frequency loop gain can be easily controlled to eliminate the impact of process errors and the high-order interface circuit can be applied to a different mechanical structure.

Based on the analytical model of ΣΔ micro-accelerometers, we derived the signal transfer function (STF) and noise transfer function (NTF) of the ΣΔ accelerometer system. The output swing of the integrators decreased when the gain of the integrators was reduced by using the proportional scaling technique. In this way, the reduction of the swing amplitude associated with the nonlinearity of the amplifier gain will lead to the reduction of the output harmonic distortion and the overall power consumption. The loop stability is ensured by controlling the zero-pole distribution of the loop filter to make sure that the average frequency response amplitude of noise transfer function is within a reasonable range. The values of feedforward coefficients, feedback coefficients and integrator gain coefficients were determined as shown in Table 3.

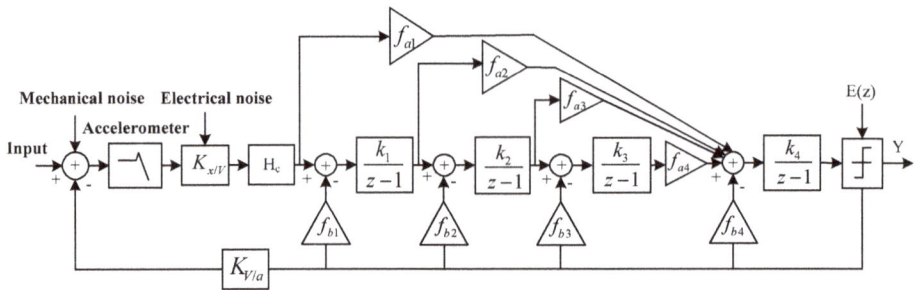

Figure 6. Analytical model of ΣΔ micro-accelerometers.

Table 3. The modulator coefficient.

Coefficient	k_1	k_2	k_3	k_4	f_{a1}	f_{a2}	f_{a3}	f_{a4}	f_{b1}	f_{b2}	f_{b3}	f_{b4}
Value	0.05	0.8	0.2	0.05	0.4	0.2	0.1	0.4	0.2	0.3	0.5	0.6

In order to stabilize the system in the high-order structure, a pre-compensator as shown in Figure 4 was added to the loop to delay the phase intersection to the gain intersection. Because the gain intersection point was very far in the high-order structure, the pre-compensator needed to provide a larger pre-phase, which required a larger compensation depth α. The increase of α will decrease the low-frequency gain, but will not affect the noise characteristics of higher-order structures. In the high-Q ΣΔ micro-accelerometers, the stability of higher order systems is strongly affected by compensation depth α. Only when α is greater than a certain critical value, the system can reach a stable state. Additionally, with the increase of Q-value, the higher order system stability requires a larger value of α. In this paper, the stability of the Sigma-Delta modulator was studied by the root locus method. The pole position of transfer function was changed by the gain of quantizer. The gain of quantizer was changed by the amplitude of the input signal. The root locus of the topology analysis model of the Sigma-Delta modulator designed is as shown in Figure 7. As the input signal amplitude increased, the quantizer gain decreased. It can be seen that from Figure 7 when the quantizer gain is more than 0.547, the root locus begins to deviate from the unit circle, which can lead to an increase in the amplitude of the input signal of the quantizer.

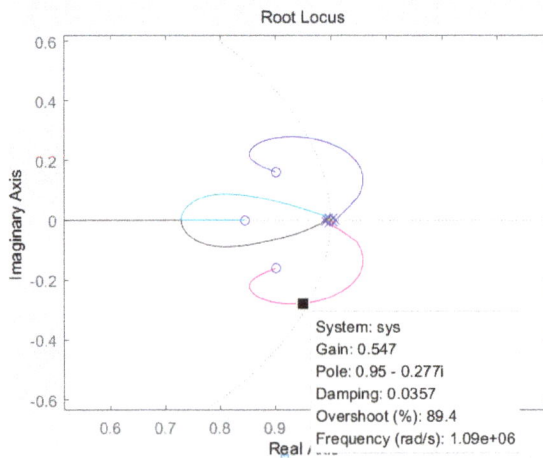

Figure 7. Root locus of the Sigma-Delta modulator.

System parameters are optimized by improving stability and reducing harmonic distortion. The reference voltage of simulation was (±2.5 V). When the sampling frequency was 250 kHz, there was an equivalent acceleration signal amplitude of 1 g and a frequency of 30.5175 Hz. Figure 8 shows the output transient waveforms of the first-stage integrator, the second-stage integrator, the third-stage integrator and the fourth-stage integrator in sequence from top to bottom. It can be seen from Figure 8 that the output amplitude of the integrators was within a very small range of ±0.2 V. It shows that the topology of the Sigma-Delta modulator designed in this paper has the advantage of small output swing and good stability.

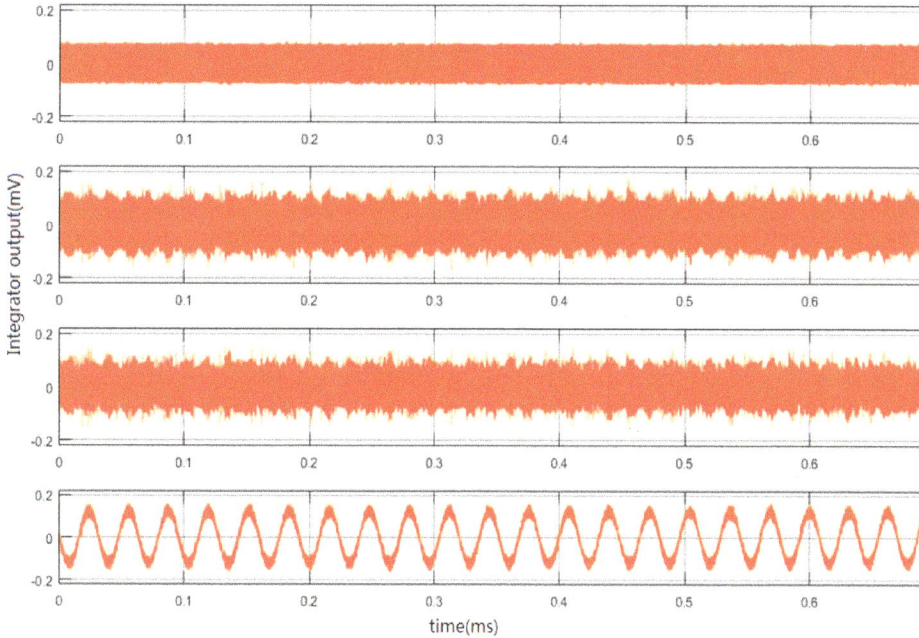

Figure 8. Output waves of each stage of the integrator.

3.2. The Test of Digital Micro-Accelerometers

The ΣΔ modulator interface circuit for micro-accelerometers was fabricated in a standard 0.35 μm four layers metal double polycrystal CMOS process and the printed circuit board (PCB) photograph of the digital micro-accelerometer system is shown in Figure 9. The photograph of the interface circuit chip is also shown in Figure 9, which has 28 pins for the chip test. The active area of the chip was 3.3 mm × 3.5 mm. The 5 V power supply of the interface circuit combined with the sensitive element was supported by the Agilent E3631 (Agilent Technologies Inc, Santa Clara, CA, USA). The input signal (240 Hz) and clock signal was supplied by the Tektronix AFG3102 function signal generator (Tek Technology Co., Shanghai, China). The 65536-point digital output sequence of ΣΔ micro-accelerometers was captured by an Agilent Logic analyzer 16804A (Agilent Technologies Inc, Santa Clara, CA, USA). The ouput digital signal is used to calculate the output power spectral density (PSD) as shown in Figure 9a by a MATLAB program (R2016a, MathWorks, Natick, MA, USA).

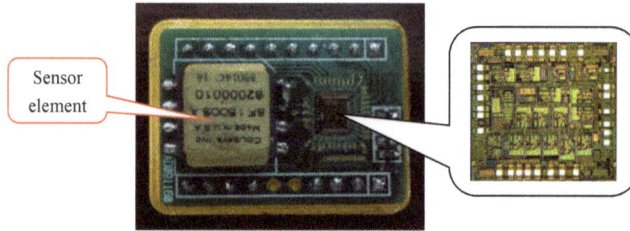

Figure 9. The printed circuit board photograph of ΣΔ modulator interface chip circuit

The power dissipation of the micro-accelerometer system was 10 mW at a sampling frequency of 250 kHz. The full scale range was ±1 g and the ΣΔ modulator had a dynamic range (DR) of 97 dB. The third harmonic distortion can be calculated by the difference between the signal-to-noise ratio of the fundamental wave and signal-to-noise ratio of the third harmonic wave in the spectrogram. The ΣΔ micro-accelerometer system can achieve a third harmonic distortion of −98 dB as shown in Figure 10a and a resulting signal-to-noise ratio (SNR) of 108 dB when referred to 1 g full scale DC acceleration. The average noise floor in low-frequency range was less than −140 dBV. The ΣΔ micro-accelerometer system could achieve a resolution of 0.48 μg/Hz$^{1/2}$ over a signal bandwidth. The test of the linearity is as shown in Figure 10b by the fitting of a straight line at ±1 g full scale. The ΣΔ micro-accelerometers could achieve a nonlinearity of 0.15% FS (full scale). After further electromagnetic shielding and vibration reduction, the output of the micro-accelerometer system was sampled when the sensor was at the state of zero acceleration in the laboratory test environment. The sampling time was longer than 4 h. After processing the sampled data with the Allen variance program in MATLAB, the bias stability test results of the closed-loop micro-accelerometer are shown as Figure 10c. The internal embedding plot in Figure 10c is processed sample data, and the bias stability is about 18 μg by calculation. We replaced 30 ASIC chips for the same sensitive structure and repeated the test. The bias stability of the closed-loop ΣΔ micro-accelerometer system was within 30 μg. Therefore, the micro-accelerometer system integrated with an ASIC chip had good output stability.

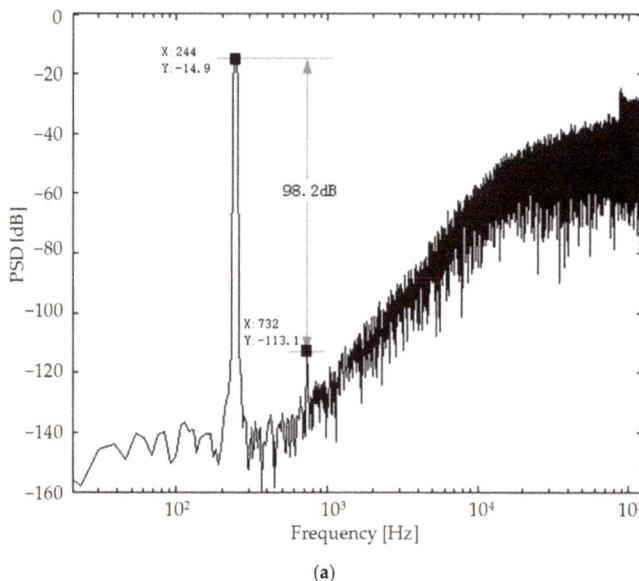

(a)

Figure 10. *Cont.*

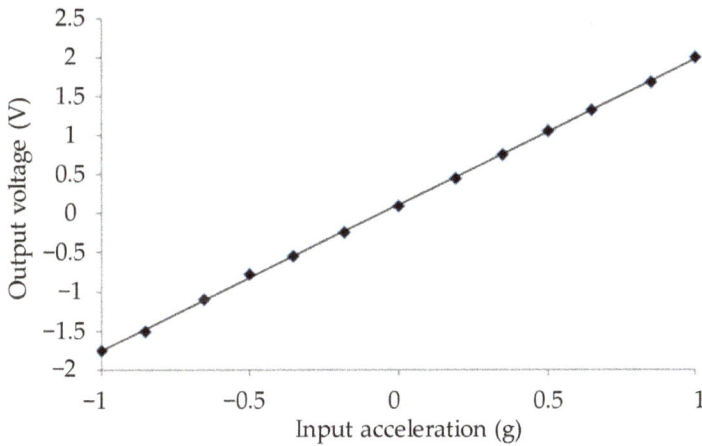

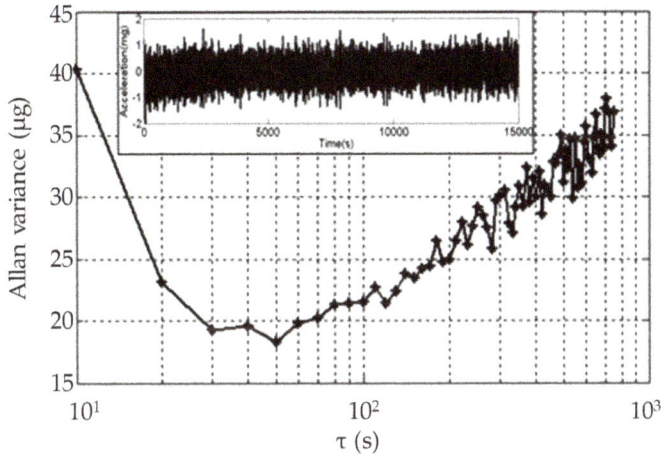

Figure 10. (a) The power spectrum density test of the digital accelerometer system; (b) the test of nonlinearity; (c) the test of bias stability.

4. Conclusions

In this work, we proposed a high-order ΣΔ high-Q micro-accelerometer. In the ΣΔ interface ASIC, we used the correlated double sampling technique to eliminate the $1/f$ noise and offset for low-noise front-end detection. Additionally, the gain of the integrators was reduced by using the proportional scaling technique. The stability of high-order ΣΔ was studied by the root locus method. The interface circuit was fabricated in a standard 0.35 μm CMOS process. The test results of the system showed that: The micro-accelerometer could achieve a signal-to-noise ratio (SNR) of 108 dB; an average noise floor in low-frequency range of less than −140 dBV and a third harmonic distortion of −98 dB; a resolution of 0.48 μg/Hz$^{1/2}$ (@300 Hz); a bias stability of 18 μg by the Allen variance program in MATLAB.

As shown in Table 2, the ΣΔ micro-accelerometer system could achieve a better performance than most of the reported accelerometers in Table 4.

Table 4. Comparison of this work with other micro-accelerometers.

Parameters	[16]	[17]	[18]	[19]	This Work
Bandwidth (Hz)	200	300	500	300	300
Sensitivity (V/g)	0.495	2.267	NA	0.373	1.866
Noise floor ($\mu g/Hz^{1/2}$)	2	0.3	4	1.15	0.48
Power (mW)	3.6	85.8	4.5	12	10
Process (μm)	0.35	0.7	0.5	0.6	0.35
Supply/Range	3.6 V/±1.15 g	5 V/±1.5 g	3 V/NA	9 V/±11 g	5 V/±1 g
Figure of Merit (FOM)	0.51	1.49	0.80	0.80	0.28

We compared our work with the previously reported accelerometers based on a representative figure of merit (FOM = $P \times a_n \times BW^{1/2}/BW$), where P is the power dissipation, a_n is the noise floor and BW is the signal bandwidth. This work is advantageous in the noise floor compared with [16,18,19] and a better FOM as shown in Table 2. We propose this interface ASIC based on the $\Sigma\Delta$ micro-accelerometer, which can satisfy the high-precision application in digital micro-accelerometers. The technical index of comprehensive performance can achieve a certain level.

Author Contributions: X.L. and J.H. designed the signal processing ASIC; X.L. designed the layout of ASIC; X.L. performed the experiments and wrote this paper.

Funding: This research was funded by [National Natural Science Foundation of China] grant number [61671259], [Zhejiang Provincial Natural Science Foundation] grant number [LY19F010005] and sponsored by K.C. Wong Magna Fund in Ningbo University. The APC was funded by [National Natural Science Foundation of China].

Acknowledgments: The authors would like to thank the National Natural Science Foundation of China (No 61671259), Zhejiang Provincial Natural Science Foundation (No. LY19F010005) and sponsored by K.C. Wong Magna Fund in Ningbo University.

Conflicts of Interest: The authors declare no conflicts of interest.

References

1. Song, Z.; Sun, T.; Wu, J. System-Level Simulation and Implementation for a High Q Capacitive Accelerometer with PD Feedback Compensation. *Microsyst. Tech.* **2014**, *21*, 2233–2240. [CrossRef]
2. Wang, Y.M.; Chan, P.K.; Li, H.K.H. A Low-Power Highly-Sensitive Capacitive Accelerometer IC Using Auto-Zero Time-Multiplexed Differential Technique. *IEEE Sens. J.* **2015**, *15*, 6179–6191. [CrossRef]
3. Paavola, M.; Kämäräinen, M.; Laulainen, E. A Micropower-Based Interface ASIC for a Capacitive 3-Axis Micro-Accelerometer. *IEEE J. Solid-State Circuits* **2009**, *44*, 3193–3210. [CrossRef]
4. Dong, Y.F.; Kraft, M.; White, W.R. Higher Order Noise-Shaping Filters for High-Performance Micromachined Accelerometers. *IEEE Trans. Instrumen. Meas* **2007**, *56*, 1666–1674. [CrossRef]
5. Bajdechi, O.; Huijsing, J.H. A 1.8-V $\Delta\Sigma$ modulator interface for an electret microphone with on-chip reference. *IEEE J. Solid-State Circuits* **2002**, *37*, 279–285. [CrossRef]
6. Petkov, V.P.; Balachandran, G.K.; Beintner, J. A Fully Differential Charge-Balanced Accelerometer for Electronic Stability Control. *IEEE J. Solid-State Circuits* **2014**, *49*, 262–270. [CrossRef]
7. Petkov, V.P.; Boser, B.E. A fourth-order $\Sigma\Delta$ interface for micro-machined inertial sensors. *IEEE J. Solid-State Circuits* **2005**, *40*, 1602–1609. [CrossRef]
8. Lajevardi, P.; Petkov, V.P.; Murmann, B. A $\Sigma\Delta$ interface for MEMS accelerometers using electrostatic spring constant modulation for cancellation of bondwire capacitance drift. *IEEE J. Solid-State Circuits* **2013**, *48*, 265–275. [CrossRef]
9. Chiang, C.-T.; Peng, C.-R. A Multi-Level Force-Feedback CTV-Based Analog Sensing Circuits with Delta-Sigma Modulator for CMOS Micro-Accelerometers. In Proceedings of the 2015 IEEE 12th International Conference on Networking, Sensing and Control, Taibei, Taiwan, 9–11 April 2015; IEEE: Piscataway, NJ, USA, 2015.
10. Huang, F.; Liang, Y. Analysis and design of the system of a total digital Si-gyroscope. *Int. J. Mod. Phys. B* **2017**, *31*, 1741008. [CrossRef]

11. Maria Gomez, J.; Bota, S.A.; Marco, S. Force-Balance Interface Circuit Based on Floating MOSFET Capacitors for Micro-Machined Capacitive Accelerometers. *IEEE Trans. Circuits Syst.* **2006**, *53*, 546–552. [CrossRef]
12. Soen, J.; Voda, A.; Condemine, C. Controller Design for a Closed-Loop Micromachined Accelerometer. *Control Eng. Pract.* **2007**, *15*, 57–68. [CrossRef]
13. Ko, H.; Cho, D. Low Noise Accelerometer Microsystem with Highly Configurable Capacitive Interface. *Analog Integr. Circuits Signal Process.* **2011**, *67*, 365–373. [CrossRef]
14. Xiang, L.; Jian, H.; Wei, C.; Xiao, L.; Liang, Y. A Novel High-Precision Digital Tunneling Magnetic Resistance-Type Sensor for the Nanosatellites' Space Application. *Micromachines* **2018**, *9*, 121–140.
15. Wu, P.C.; Liu, B.D.; Yeh, C.Y. Design of a 0.6-V 0.2-mW CMOS MEMS Accelerometer. In Proceedings of the 2015 IEEE International Conference on Consumer Electronics—Taiwan, Taipei, Taiwan, 6–8 June 2015; IEEE: Piscataway, NJ, USA, 2015.
16. Yucetas, M.; Pulkkinen, M.; Kalanti, A. A high-resolution accelerometer with electrostatic damping and improved supply sensitivity. *IEEE J. Sens.* **2012**, *47*, 1721–1730. [CrossRef]
17. Aaltonen, L.; Halonen, K. Continuous-time interface for a micromachined capacitive accelerometer with NEA of 4 g and bandwidth of 300 Hz. *Sens. Actuators A* **2009**, *154*, 46–56. [CrossRef]
18. Amini, B.V.; Abdolvand, R.; Ayazi, F. A 4.5-mW closed-loop micro-gravity CMOS SOI accelerometer. *IEEE J. Solid-State Circuits* **2006**, *41*, 2983–2991. [CrossRef]
19. Pastre, M.; Kayal, M.; Schmid, H.; Huber, A. A 300 Hz 19 b DR capacitive accelerometer based on a versatile front end in a 5th-order ΔΣ loop. In Proceedings of the 2009 Proceedings of ESSCIRC, Athens, Greece, 14–18 September 2009; IEEE: Piscataway, NJ, USA, 2009.

micromachines

MDPI

Article

High Performance Seesaw Torsional CMOS-MEMS Relay Using Tungsten VIA Layer

Martín Riverola, Francesc Torres, Arantxa Uranga and Núria Barniol *

Department of Electronics Engineering, Universitat Autònoma de Barcelona, 08193 Bellaterra, Spain;
Martin.riverola@gmail.com (M.R.); francesc.torres@uab.es (F.T.); arantxa.uranga@uab.es (A.U.)
* Correspondence: nuria.barniol@uab.cat; Tel.: +34-93-581-1361

Received: 28 September 2018; Accepted: 1 November 2018; Published: 7 November 2018

Abstract: In this paper, a seesaw torsional relay monolithically integrated in a standard 0.35 μm complementary metal oxide semiconductor (CMOS) technology is presented. The seesaw relay is fabricated using the Back-End-Of-Line (BEOL) layers available, specifically using the tungsten VIA3 layer of a 0.35 μm CMOS technology. Three different contact materials are studied to discriminate which is the most adequate as a mechanical relay. The robustness of the relay is proved, and its main characteristics as a relay for the three different contact interfaces are provided. The seesaw relay is capable of a double hysteretic switching cycle, providing compactness for mechanical logic processing. The low contact resistance achieved with the TiN/W mechanical contact with high cycling life time is competitive in comparison with the state-of-the art.

Keywords: MEMS relays; MEMS switches; mechanical relays; CMOS-MEMS; MEMS

1. Introduction

It is expected that new micro- and nanoelectromechanical (M/NEM) relays can play an important role as a new device for adding functionality and decreasing the power consumption for the more demanding area of consumable devices (IoT, wearables) [1]. One of the important things in mechanical relays is the capability of a quasi-ideal switching behavior (with a very abrupt on-off switching, and zero current leakage during the OFF-state) and multi-terminal operation which can serve to save energy, as it has been envisioned in several different digital applications [2–5]. The possibility of using the complementary metal oxide semiconductor (CMOS) platform for the monolithic fabrication of such M/NEMS relays in a real combination with classical CMOS devices can open a myriad of new possibilities for decreasing power consumption. Additionally, the high number of metal layers used in the advanced CMOS technology nodes make very attractive the exploitation of a CMOS-MEMS platform for using metal layers, not only as an electrical connection path, but also to provide some active processing using these layers as embedded MEMS devices [6,7]. Despite this interest in obtaining functional mechanical switching devices embedded in CMOS, most of the presented examples from the literature are only CMOS-compatible [8–12], with few of them being really embedded in CMOS [13–17]. In all cases, the devices are far from possessing all of the ideal characteristics (low contact resistance, low operation voltage and high yield). For instance, the TiN coated relay presented in [8] presents a non-ohmic contact resistance with a high life cycling, while the similarly TiN coated PolySilicon relay in [9] has low contact resistance, but presents limited cycling operation. In Reference [10], a NEMS relay with a very low pull-in voltage (0.4 V) is presented, but it is only operable for 20 cycles. In Reference [11], a demonstration of a CMOS driven Pt-NEMS relay fabricated over the CMOS is presented, but with a very high contact resistance (100 MΩ) and without testing the life time of the relay. Reference [12] presents a two-terminal TiN NEMS relay fabricated under a CMOS compatible process with an operability of hundreds of cycles, but with a limited current operation (nA range). Concerning papers with MEMS relays embedded in CMOS, similar problems are encountered. Papers using the same CMOS-MEMS tungsten-based relay as presented in this paper, but with different configurations and designs, suffers from these non-ideal characteristics: Reference [13] presents a torsional relay with a high pull-in voltage and below one hundred operation cycling; References [14,15] are based on lateral relays exhibiting in both cases a high contact resistance (1 MΩ and 750 MΩ in References [14,15], respectively). Even higher contact resistances and low cycling operation are encountered in other CMOS-MEMS approaches: In Reference [16], contact resistance is greater than 500 MΩ and 30 operation cycles; in Reference [17], the contact resistance is in the GΩ range and only 10 operation cycles. As a consequence of these reported characteristics, more research is necessary in order to improve the performance of these CMOS-MEMS relays.

In this paper, we present new MEMS devices capable of providing five-terminal relays with a bidirectional operation and embedded in CMOS, demonstrating enhanced performance compared with the already reported TiN-based MEMS relays. The main issue with the fabrication of the presented relays is the use of the tungsten VIA of the conventional AMS (Austria Microsystems) 0.35 μm CMOS technology. The exploitation of the VIA3 made from tungsten as the main structural layer for CMOS-MEMS devices presents a series of attractive characteristics that are suitable for mechanical relays: high hardness, being resistant to wear and plastic deformation; high melting point (tungsten exhibits the highest melting point); being resistant to welding-induced failure due to Joule heating at the contact. Furthermore, VIA3 is a top Back-End-Of-Line (BEOL) layer more thinly covered in SiO_2, which implies small releasing times, and thus increased yield in the fabrication process.

The use of the tungsten VIA3 has been demonstrated previously for MEMS devices: resonators for monolithically CMOS-MEMS stand-alone oscillators [18,19], relays for switching applications [13–15], and very recently as CO_2 transducers [20]. All these applications demonstrate the importance of the approach and the opportunity to explore new MEMS structures and devices based on this tungsten VIA3 approach. In this paper we will focus on a mechanical five-terminal relay working in its torsional operation with an enhancement of the electrostatic coupling, and consequently lower pull-in voltage, and a decrease of the contact resistance due to the ability to define larger contact areas compared with the above reported examples. Moreover, the paper studies all the different contact materials available in the BEOL-CMOS metal layers without adding any additional metallization in order to provide a totally monolithic integration with CMOS. From the presented results we can state that the CMOS-MEMS relay with TiN-W contacts presents the highest ON-OFF current ratio (10^7), the lowest contact resistance 2 kΩ, and the highest cycling life test compared with the state-of-the-art MEMS relays based on TiN contacts [8,9,12–17].

2. Materials and Methods

2.1. Device Design and Fabrication

The torsional relay consists of a five-terminal seesaw device schematically drawn in Figure 1. The seesaw relay design consists in a main plate formed by two sandwiched metal layers (MET3 and MET4) of the CMOS technology contacted through the contacting metal VIA layer (specifically, a sandwiched MET4-VIA3-MET3). This main plate is anchored by two VIA3 torsional beams (called source, S) which allow the ends of the main beam to move up and down by electrostatically actuating the relay with the basally located gate electrodes (G_R and G_L). This gate electrode is formed by metal layer (MET1) and its contacting VIA (VIA1). Three types of endings (the final contacts for the seesaw relays) are made (see cross-section A3–A4 in Figure 1c): (a) Type I, MET4-VIA3-MET3, which make contact with the drain electrodes made by MET2, (b) Type II, MET4-VIA3, which make contact with the drain electrodes of MET2; and (c) Type III, MET4-VIA3, which makes contact with the drain electrodes defined in this case with MET2-VIA2. Each of the metal layers (METi) of the 0.35 μm CMOS technology are a sandwiched layer consisting of TiN/Al/TiN. In this sense, three kinds of contacts will be characterized: (a) TiN vs. TiN in type I relays; (b) W vs. TiN in type II relays;

and (c) W vs. W in type III relays. Note that these three types of relays will provide contact gaps at different heights.

The design parameter values for the three types of relays are listed in Table 1. The parameters used have been chosen taking into account the following requirements: (a) torsional actuation selecting VIA3 torsional beams to have an equivalent torsional spring constant smaller than the vertical actuation, using the minima dimension for the VIA3 width ($W_T = 0.5$ μm), and gate electrodes (G_R and G_L) are situated at the end of the body to promote torsional movement; (b) maximize actuation area (gate electrodes size and body size) between MET3 and MET1 to minimize pull-in voltage in comparison with previous designs [13] (note that the VIA1 contacts used over the MET1 are intended to enhance electrostatic coupling between actuation electrodes and relay body to further reduce pull-in voltage); (c) squared contact area of 2.5 μm × 2.5 μm to decrease contact resistance. All of the other parameters are constraints from the CMOS technology used. Due to the non-uniform material based seesaw relay, as well as to the structure of the gate electrodes (with the small metal contacts, VIA1), it is not possible to analytically compute the behavior of the seesaw relay (i.e., pull-in voltage). Consequently, finite-element-model simulations using Coventor have been extensively used to tune design parameters (Table 2 summarizes the main simulated characteristics for the seesaw relays).

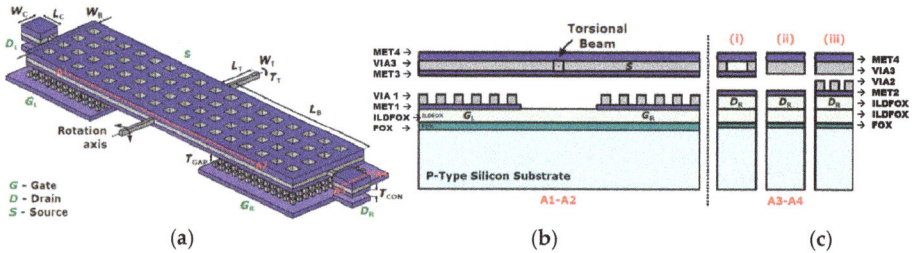

Figure 1. (**a**) 3D schematic of the designed seesaw relay including cross-sectional views (red lines). (**b**) Cross-section A1–A2 along the length of the relay, the gate electrodes are defined with MET1 and VIA1. (**c**) Cross-section A3–A4 at the contact area of the relay (between Source and Drain) with the three possibilities: (i) Type I, (ii) Type II, and (iii) Type III.

Table 1. Seesaw relay design parameters and their values.

Design Parameter	Value (μm)
Torsion beam length L_T	4.7
Torsion beam width W_T	0.5
Torsion beam thickness T_T	1.3
Body length L_B	59.6
Body width W_B	16
Electrode length $G_{R,L}$	30
Electrode width G_W	16
Contact length L_C	2.5
Contact width W_C	2.5
Actuation gap T_{Gap}	1.95
Contact gap (i) T_{con}	1
Contact gap (ii) T_{con}	1.3 [a]
Contact gap (iii) T_{con}	0.45 [a]

[a] The gap is measured after fabrication.

The fabrication process of the VIA3 MEMS structures is based on a mask-less wet-etching process [21,22]. A passivation aperture is defined over the resonator which allows this in-house post-CMOS MEMS releasing process to be done directly while the passivation layer of silicon nitride is used as a protective layer for the rest of the chip. The releasing process consists basically of three steps: (a) isotropic wet-etching in a bath of buffered hydrofluoric acid solution at room temperature (with an oxide etching rate of around 300 nm/min [21]); (b) chip washing in distilled water followed by an isopropyl alcohol bath to eliminate the water; and (c) heating in an oven at 100 °C to evaporate the remaining alcohol. No sticking problems have been encountered for the seesaw MEMS relay with this etching, which does not require critical point drying for the releasing. As it is an isotropic process, the etching time depends on the MEMS dimensions and the quantity of oxide over the structure. In the case of the seesaw relays, and due to the large area of the body structure, releasing holes have been included to facilitate the wet etching of the sacrificial SiO_2 layer underneath the large main plate. The etching time used was typically in the range between 10 and 18 min. This etching process is CMOS compatible, as it has already been demonstrated with VIA3 MEMS structures embedded in functional CMOS circuitry [18,19,23].

It is necessary to ensure that the torsional mode operation of the seesaw relay dominates over the flexural mode operation while it is switching. Therefore, the vertical flexural spring constant must be much stiffer than the torsional spring constant. Table 2 shows the simulated resonant frequency of the torsional and vertical mode and their respective effective stiffness using the following material properties: Young modulus of 410 GPa, 70 GPa and 600 GPa, and mass densities of 19,300 kg/m^3, 2700 kg/m^3 and 5430 kg/m^3 for tungsten, aluminum and titanium nitride, respectively. As can be seen, the vertical spring constant is 55× higher than the torsional spring constant.

Figures 2 and 3 show the layout, optical and SEM images of the fabricated seesaw relays, along with the focused ion beam (FIB) cross-sectional views to detail the different technological implementations of the relay body (Figure 2) and relay contact (Figure 3). The cross-sections are provided before and after the releasing of the seesaw relay. From these images, the gap distances of the relay contact (Table 1) are extracted.

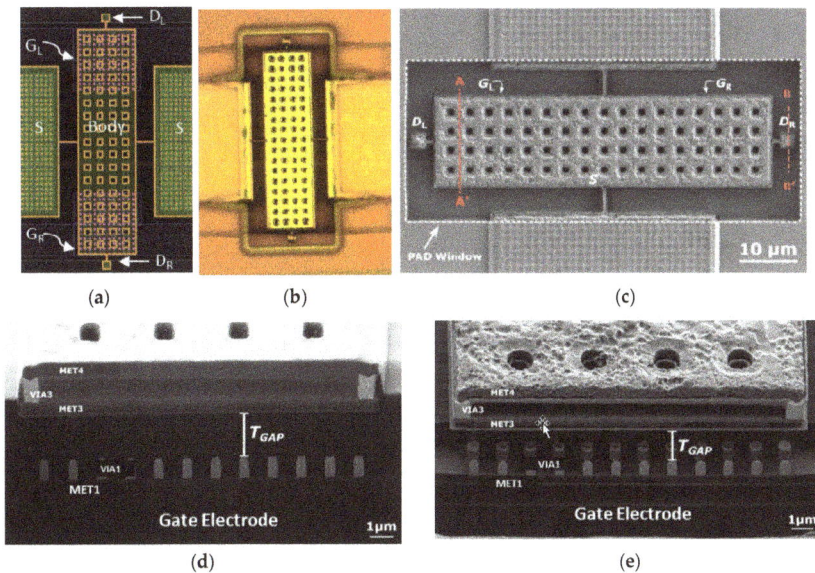

Figure 2. (**a**) Layout of the seesaw relay. The gate electrodes are the pink areas below the body structure. (**b**) Top view optical image of fabricated and released seesaw relay. (**c**) Top view SEM image indicating the cut-line A-A′ over the body structure and B-B′ over the contact area. (**d**,**e**) SEM images of the cross-section in the A-A′ cut-line (**d**) before and (**e**) after the releasing process. These images allow one to see the gate electrodes composed by the MET1 and VIA1 layers, as well as the sandwiched composition of the body element of the relay (a sandwich of MET3-VIA3-MET4).

Before Etching **After Etching**

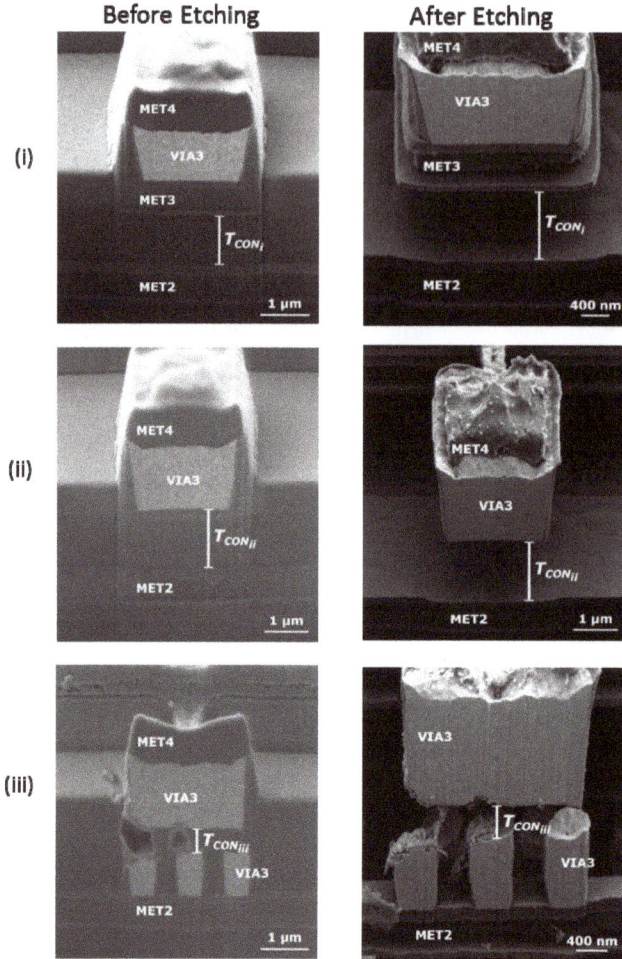

Figure 3. SEM images of the cross-section B-B' in Figure 2c over the contact in the three different designs (contact between body relay with different composition and drain) showing before (left images) and after (right images) the releasing: (**i**) Type I, (**ii**) Type II and (**iii**) Type III. These figures can be compared with the cross-section A3-A4 at the contact area of the relay in Figure 1, in which the different composition for contact source and drain are explained.

Table 2. CoventorWare simulation of the resonant frequencies and mode shapes of the seesaw relay.

	Torsional Mode	Vertical Mode
Mode Shape		
Resonant Frequency, f_0	152 kHz	1.5 MHz
Spring Constant, k_{eff}	1.28 N/m	69.6 N/m

2.2. Electrical Characterization

The fabricated seesaw relays were tested under two different conditions: (1) at room temperature in air at atmospheric pressure, and (2) under vacuum at 10^{-5} mbar. In ambient conditions, the chips were exposed to air and tested in a Cascade Microtech probe station (PM8). Under vacuum conditions, the chip was mounted and bounded onto a printed circuit boardand placed inside a homemade vacuum chamber. The current-voltage (*I-V*) characterization was performed with an Agilent semiconductor analyzer B1500A equipped with four high-resolution source measure units (SMU) (Figure 4).

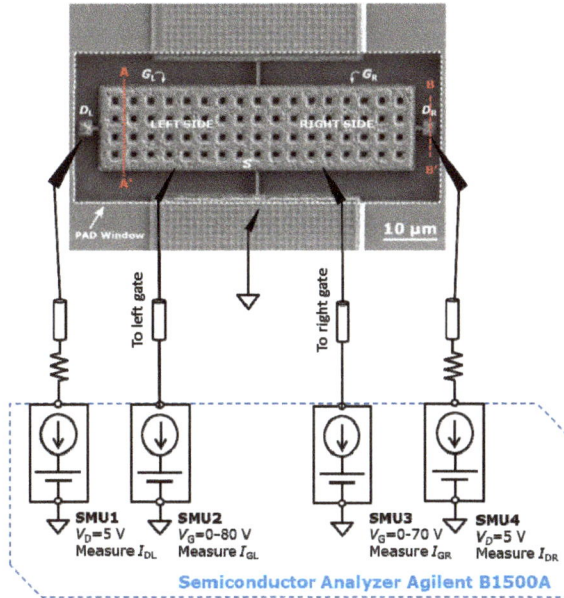

Figure 4. Electrical set-up for the current voltage (*I-V*) switching characteristics of the five-terminal relay. Four high-resolution source measure units (SMU) are used: the source electrode (relay structure) is grounded, drain electrodes (left and right) are fixed to a VD = 5 V, gate electrodes (left and right) are swept up and down from 0 to a voltage gate V_G voltage higher than the pull-in voltage, V_{PI}. Note that gates and drains are underneath the relay structure and are not visible in the image.

3. Results

In this section, the current voltage (*I-V*) curves for the three types of fabricated seesaw relays placed in both air conditions and vacuum conditions are reported. The pull-in and pull-out voltages, I_{ON}-I_{OFF} ratio, contact resistance, and the cycling, or life-time, of the different relays are provided.

3.1. Seesaw Relay with Contact Type I: TiN vs. TiN

Figure 5a,b shows the first nine current voltage (*I-V*) curves taken from both the left and right ends of a seesaw relay being exposed to air conditions. As the right gate voltage V_{GR} is increased from 0 to 85 V, the right side of the torsion beam turns on abruptly at 54.8 V, while the left side remains off. Thus, a conductive path is formed between the right contact electrode (or right drain) and the movable structure (or source) by fixing the drain-to-source voltage (V_{DS}) to 5 V. Similarly, the left side of the relay is also actuated by sweeping up and down the left gate voltage V_{GL} from 0 to 85 V and fixing the left drain voltage V_{DL} also to 5 V (protected with 1 MΩ). In this case, the left side turns on abruptly at 55.5 V. For both sweeps, the measured on-off current ratio is ~10^5, and the contact resistance R_c is ~10^8. Instead, an asymmetric behavior is observed comparing the V_{PO} of both tested sides. Since the V_{PO}, and thus the hysteresis window, is strongly related with the adhesion forces at the contact interface, this would mean that different contact scenarios are involved in both contact ends. SEM images were taken to confirm this hypothesis, as shown in Figure 6. As can be seen, the bottom thin TiN layer that forms the sandwiched MET3 layer of TiN-Al-TiN fell over the MET2 layer due to the long wet-etching to release the structure, causing the observed asymmetry in the hysteresis window.

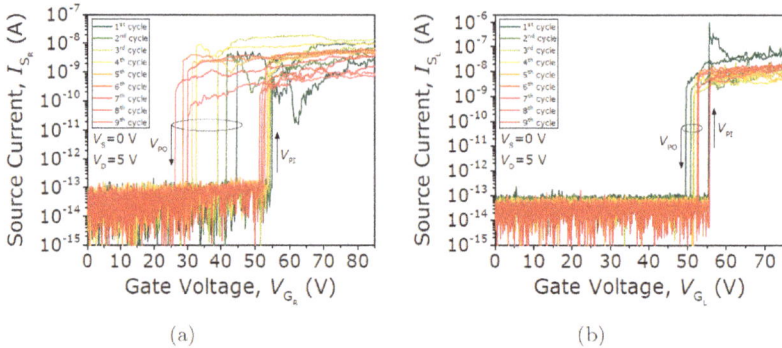

(a)   (b)

Figure 5. First nine current voltage (*I-V*) switching characteristics in ambient conditions of the (**a**) left and (**b**) right drain electrodes for the seesaw relay Type I (TiN-TiN contact).

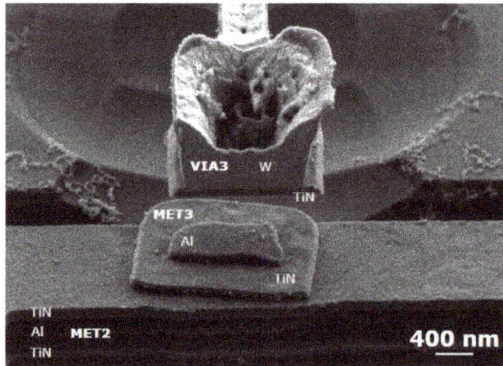

Figure 6. SEM image taken in the contact region of the seesaw relay showing the over-etch of the Al layer contained in the sandwiched MET3 layer of TiN-Al-TiN.

3.2. Seesaw Relay with Contact Type II: W vs. TiN

Figure 7a shows the first ten current voltage (*I-V*) curves taken in a contact-type-(ii) seesaw relay being exposed to air conditions, exhibiting a similar Rc of ~10^8 and an I_{ON}/I_{OFF} ratio of 10^4. Figure 7b shows how V_{PI} and V_{PO} evolve over these ten cycles. V_{PI} is fairly stable, but V_{PO} increases gradually with exposure to air. This phenomenon can be explained by the reduced surface adhesive force from metallic surfaces to oxide surfaces. Therefore, the hysteresis window reduces over time due to oxide formation in the W surface. Figure 8 shows the I-V characterization conducted under vacuum conditions at 10^{-4} mbar. The first current voltage (*I-V*) curve shows no abrupt transition due to the breakdown of the native oxide at the TiN/W contact interface (see Figure 8a). Next, ten current voltage (*I-V*) curves are taken as shown in Figure 8b, which already show the typical hysteretic behavior with initial sharp V_{PI} and V_{PO} voltages of 57.4 V and 14.6 V, respectively. The R_C is ~1 MΩ, 500× better compared to air conditions, which leads to an increased I_{ON}/I_{OFF} ratio of 10^7. Recall that a wider hysteresis window means that adhesion forces are exacerbated in the contacting region due to an increased effective contact area from the larger levels of current obtained.

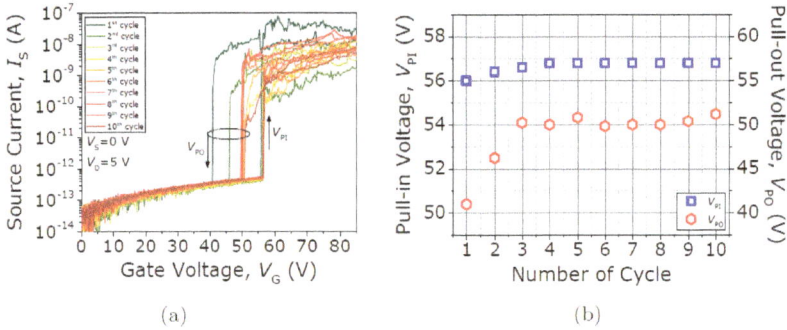

(a)   (b)

Figure 7. (**a**) First ten current voltage (*I-V*) switching characteristics. (**b**) Evolution of V_{PI} and V_{PO} over these ten cycles. Measures correspond to the seesaw relay Type II (W-TiN contact) in ambient conditions.

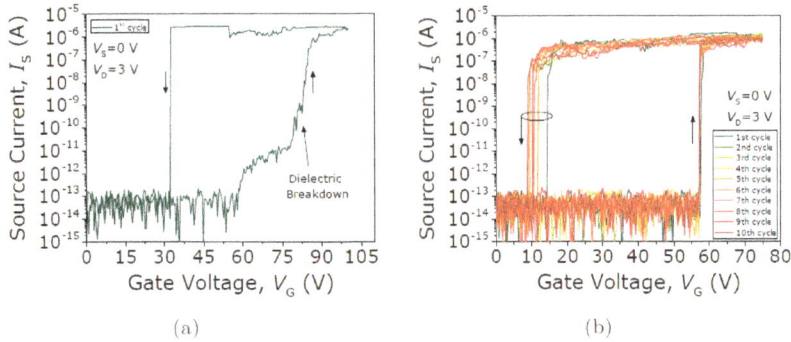

(a)   (b)

Figure 8. (**a**) First taken current voltage (*I-V*) curve showing no abrupt transition during the pull-in until the breakdown of the native oxide. (**b**) Next 10 current voltage (*I-V*) curves. Measures correspond to the seesaw relay Type II (W-TiN contact) under vacuum conditions.

Figure 9 shows how V_{PI}, V_{PO} and R_c evolve over a total of 355 switching cycles. Compliance was set over the maximum level of measured current. A nominal V_{PI} of 57 V is found to be stable over these cycles, with an absolute error of only 0.75 V. V_{PO} appears to increase over these cycles. Unexpectedly, it was found that R_c drops to 2 kΩ from cycle 251, ultimately leading to permanent stiction. This effect can be due to excessive localized Joule heating at the contact asperities, which at sufficient contact temperature, annealing of the contact takes place, reducing the contact hardness. The final 2 kΩ contact resistance is the smallest R_C found.

The V_{PI}, V_{PO} and R_c are recorded over 200 cycles in a new fresh relay (Figure 10), but this time keeping the compliance limit to 1 μA to avoid excessive Joule heating. The V_{PI} shows a nominal value of 58.2 V, with an absolute error of only 0.4 V over these cycles, demonstrating again the great stability of the VIA3 platform. Regarding the R_c, it is shown to increase with the switching cycles. Therefore, the compliance limit at 1 mA favors avoiding excessive Joule heating, but favors the insulating native-oxide formation at the contacting interface (W site of the relay), increasing the R_c. To substantiate this, Figure 11 shows the acquired current with the relay in the ON-state (V_G = 75 V >> V_{PI}), applying higher V_{DS} voltages (V_{DS} > 3 V); the current level is higher for higher V_{DS} after breaking down the grown oxide, restoring the contact performance. This indicates that the contact endurance is not intrinsically degraded but strongly affected by the oxide regrowth.

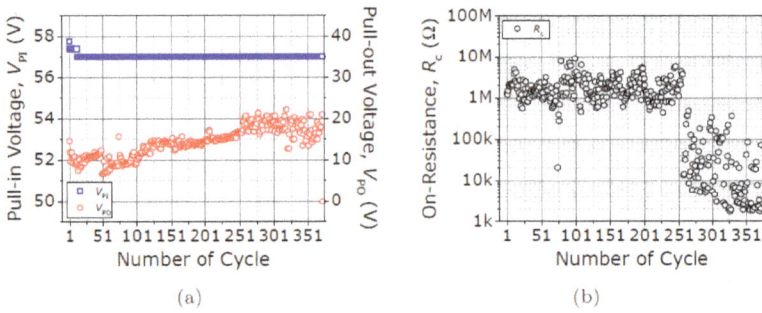

Figure 9. Evolution of (**a**) V_{PI}, V_{PO} and (**b**) R_c over 355 switching cycles under vacuum conditions and V_{DS} fixed at 3 V. Compliance current set over the measured current level.

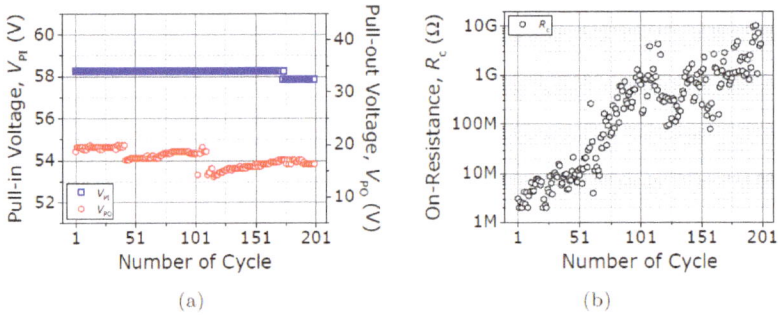

Figure 10. Evolution of (**a**) V_{PI}, V_{PO} and (**b**) R_c over 200 switching cycles under vacuum conditions and V_{DS} fixed at 3 V. Compliance current limit is fixed to 1 μA.

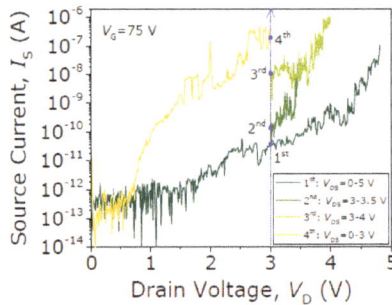

Figure 11. Successive source current versus drain voltage (I_{DS}-V_{DS}) sweeps showing the restoring of the contact performance after breaking down the grown native oxide on the W contact site of the relay. V_{GS} is fixed to 75 V to keep the relay in the ON-state.

3.3. Seesaw Relay with Contact Type III: W vs. W

Figure 12 shows the I-V characterization of both left and right ends of a contact Type III seesaw relay being exposed to air conditions. It can be observed an initial symmetric V_{PI} of 47.4 and 47.1 V in the left and right ends respectively. However, the current degrades to the noise level in only five cycles. Thus, contact Type III seesaw involving W-to-W interfaces exhibit the most exacerbated degradation when cycling in air.

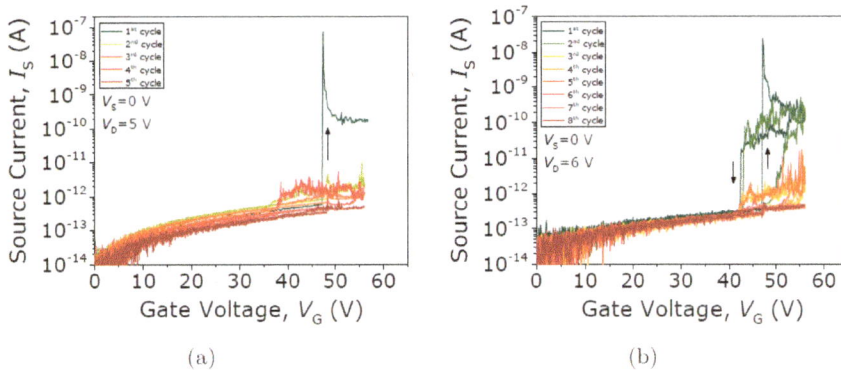

Figure 12. Successive current voltage (*I-V*) curves in air from the (**a**) left contact and (**b**) right contact of the Type III seesaw relay.

In contrast, the contact performance of the Type III seesaw relay is found to behave completely differently when it is operated under vacuum conditions. First, the initial native oxide breakdown is produced switching the device on (with V_{GS} = 75 V) and sweeping up the V_{DS} until the drain current spike is detected (see Figure 13). After this non-conductive oxide breakdown, I-V characteristics of the same relay for four different V_{DS} bias voltages are acquired (see Figure 14). The same pull-in and pull-out voltages are obtained no matter the V_{DS} bias used, as expected. Only the level of current in the ON-state is changed according with the V_{DS} bias. In Figure 14b, the R_C is computed sweeping the V_{DS} voltage while the relay is in its ON-state (V_{GS} = 75 V), obtaining a value of 51.4 kΩ. An attempt is then made to monitor the evolution of contact properties after each cycle by taking continuous I-V curves with fixed V_{DS} = 1 V (Figure 15). By doing so, V_{PI} is found to be stable, but the relay is stuck after 16 cycles, which indicates a lower cycling life compared with the Type II seesaw relays. In Table 3, a brief summary of the three types of relays based on the seesaw torsional structure is provided.

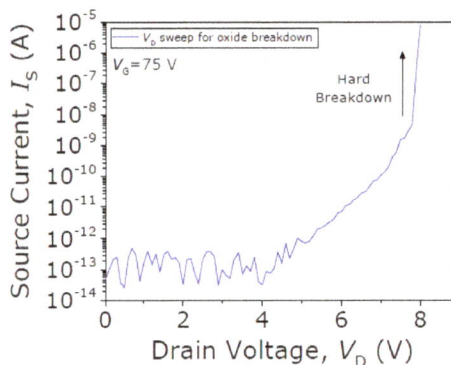

Figure 13. Current versus drain voltage curve in pull-in conditions to produce the initial native oxide breakdown (roughly at V_{DS} = 8 V).

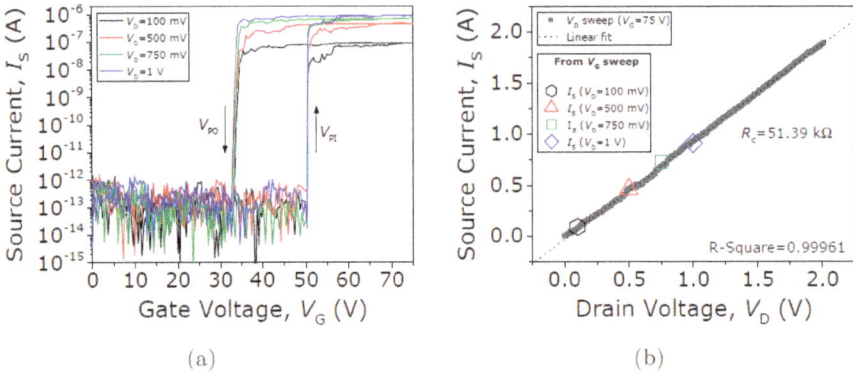

Figure 14. (**a**) Current voltage (*I-V*) curves for different V_D voltages. (**b**) I_{DS}-V_{DS} sweep with the relay in its ON-state (V_G = 75 V), showing an ohmic dependence. The R_C values extracted from the *I-V* curves in (**a**) are also plotted (symbols).

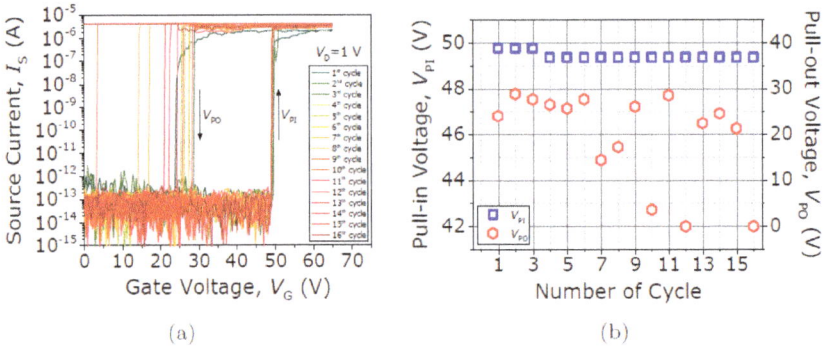

Figure 15. (**a**) Current voltage (*I-V*) curves for different V_D voltages. (**b**) Evolution of V_{PI} and V_{PO} over 16 cycles.

Table 3. Summary of the characterized parameters of the torsional MEMS relays integrated monolithically using the VIA3 tungsten of the BEOL layer of the CMOS 0.35 µm technology.

Device	SEM Image Structural/Material	Contact Material	V_{PI} (V)	ON-OFF Ratio	R_C (Ω)	Cycles (EOL)
This work		TiN/TiN	54.8 [a]	10^5 [a]	100 M [a]	-
		W/TiN	57	10^7	2 k	355
	Via3 BEOL-CMOS layers	W/W	49	10^7	51.4 k	16
[13]	Via3 BEOL-CMOS layers	TiN/TiN	71.3	10^6	20–45 k	65
[16]	CVD/PVD TiN, CMOS	TiN/TiN (annealed)	24	10^5	500 M	30
[8]	TiN coated PolySiGe, no CMOS	TiN/SiGe	15	10^6	No ohmic contact	10^{10}
[17]	MIM module, CMOS	TiN/TiN	5–20	10^4	1 G	10
[9]	TiN coated PolySi, no CMOS	TiN/TiN	22	$>10^4$	4–15 k	150

[a] Only measurements in air conditions due to the aluminum over-etching problem.

4. Discussion and Conclusions

From the above characterization of the performance of the relays, we can state that a symmetric switching operation with a five-terminal torsional relay has been achieved, providing lower pull-in voltage and contact resistance (Type II, W/TiN contact) than previously reported, based on the same technological VIA3 platform [13] (see Table 3). In addition, if we compare the presented five-terminal torsional Seesaw device with relays already reported with TiN contacts, the seesaw relays provide the lowest contact resistance with higher cycling time. Only Reference [9] provides similar contact resistance, but they report lower switching life time and additionally these relays are not monolithically integrated into CMOS. The life time of the presented relay could be improved through the use of a proper vacuum packaging.

Overall, the five-terminal relay allows for the operation as two independent relays (left and right contact), with the guarantee that they will never be ON at the same time—one clear advantage over the CMOS transistor-based relays. This implies that a higher degree of compactness for mechanical digital logic circuits can be achieved. In this sense, the presented device is an advancement towards more robust and reliable mechanical relays which can provide a decrease in power consumption for portable and wearable devices.

Author Contributions: Conceptualization, M.R., N.B.; fabrication and physical characterization, M.R., F.T.; electrical characterization, M.R., A.U.; writing and original draft preparation, M.R., N.B.; writing, reviewing, and editing, N.B.

Funding: This research was funded by the Spanish Government and the European Union, grant number TEC2015-66337-R (MINECO/FEDER).

Conflicts of Interest: The authors declare no conflict of interest.

References

1. Pott, V.; Kam, H.; Nathanael, R.; Jeon, J.; Alon, E.; King Liu, T.-J. Mechanical computing redux: Relays for integrated circuit applications. *Proc. IEEE* **2010**, *98*, 2076–2094. [CrossRef]
2. King Liu, T.-J.; Xu, N.; Chen, I.-R.; Qian, C.; Fujiki, J. NEM relay design for compact, ultra-low-power digital logic circuits. *Proc. IEDM* **2014**, *13*, 1–4.
3. Spencer, M.; Chen, F.; Wang, C.C.; Nathanael, R.; Fariborzi, H.; Gupta, A.; Kam, H.; Pott, V.; Jeon, J.; King Liu, T.-J.; et al. Demonstration of integrated microelectro-mechanical relay circuits for VLSI applications. *IEEE J. Solid-State Circuits* **2011**, *46*, 308–320. [CrossRef]
4. Qian, C.; Peschot, A.; Osoba, B.; Ye, Z.A.; King Liu, T.-J. Sub-100 mV computing with electro-mechanical relays. *IEEE Trans. Electron Devices* **2017**, *64*, 1323–1329. [CrossRef]
5. Lee, D.; Lee, W.S.; Chen, C.; Fallah, F.; Provine, J.; Chong, S.; Watkins, J.; Wong, H.-S.P.; Howe, R.T.; Mitra, S. Combinatorial logic design using six-terminal NEM relays. *IEEE Trans. Comput.-Aided Des.* **2013**, *32*, 655–666.
6. King Liu, T.-J.; Sikder, U.; Kato, K.; Stojanovic, V. There is plenty of room at the top. In Proceedings of the IEEE International Conference on Micro Electro Mechanical Systems, Las Vegas, NA, USA, 22–26 January 2017; pp. 2–5.
7. Xu, N.; Sun, J.; Chen, I.-R.; Hutin, L.; Chen, Y.; Fujiki, J.; Qian, C.; King Liu, T.-J. Hybrid CMOS/BEOL-NEMS technology for ultra-low-power IC applications. *Proc. IEDM* **2014**, *28*, 1–4.
8. Ramezani, M.; Severi, S.; Moussa, A.; Osman, H.; Harrie Tilmans, A.C.; De Meyer, K. Contact reliability improvement of a poly-SiGe based nano-relay with titanium nitride coating. In Proceedings of the 18th International Conference on Solid-State Sensors, Actuators and Microsystems (TRANSDUCERS), Anchorage, AK, USA, 21–25 June 2015; pp. 576–579.
9. Shavezipur, M.; Lee, W.S.; Harrinson, K.L.; Provine, J.; Mitra, S.; Wong, H.-S.P.; Howe, R.T. Laterally actuated nanoelectromechanical relays with compliant, low resistance contact. In Proceedings of the IEEE 26th International Conference on Micro Electro Mechanical Systems (MEMS), Taipei, Taiwan, 20–24 January 2013; pp. 520–523.
10. Lee, J.O.; Song, Y.H.; Kim, M.W.; Kang, M.H.; Oh, J.S.; Yang, H.H.; Yoon, J.B. A sub-1-volt nanoelectromechanical switching device. *Nat. Nanotechnol.* **2013**, *8*, 36. [CrossRef] [PubMed]

11. Chong, S.; Lee, B.; Parizi, K.B.; Provine, J.; Mitra, S.; Howe, R.T.; Wong, H.S.P. Integration of Nanoelectromechanical (NEM) Relays with Silicon CMOS with Functional CMOS-NEM Circuit. In Proceedings of the International Electron Devices Meeting, Washington, DC, USA, 5–7 December 2011.

12. Jang, W.W.; Lee, J.O.; Yoon, J.B.; Kim, M.S.; Lee, J.M.; Kim, S.M.; Cho, K.H.; Kim, D.W.; Park, D.; Lee, W.S. Fabrication and Characterization of a nanoelectromechanical switch with 15 nm thick suspension air gap. *Appl. Phys. Lett.* **2008**, *92*, 103110. [CrossRef]

13. Riverola, M.; Sobreviela, G.; Torres, F.; Uranga, A.; Barniol, N. A monolithically integrated torsional CMOS-MEMS relay. *J. Micromech. Microeng.* **2016**, *26*, 115012. [CrossRef]

14. Riverola, M.; Uranga, A.; Torres, F.; Barniol, N. Fabrication and characterization of a hammer-shaped CMOS/BEOL-embedded nanoelectromechanical (NEM) relay. *Microelectron. Eng.* **2018**, *92*, 44–51. [CrossRef]

15. Riverola, M.; Vidal-Álvarez, G.; Sobreviela, G.; Uranga, A.; Torres, F.; Barniol, N. Dynamic Properties of Three-Terminal Tungsten CMOS-NEM Relays Under Nonlinear Tapping Mode. *IEEE Sens. J.* **2016**, *16*, 5283–5291. [CrossRef]

16. Lee, J.-O.; Kim, M.-W.; Ko, S.-D.; Yoon, J.-B. Annealing effect on contact characteristics in TiN based 3-terminal NEM relays. In Proceedings of the 10th IEEE International Conference on Nanotechnology, Seoul, South Korea, 17–20 August 2010; pp. 258–261.

17. Muñoz-Gamarra, J.L.; Uranga, A.; Barniol, N. Nanomechanical switches based on metal-insulator-metal capacitors from standard complementary-metal-oxide semiconductor technology. *Appl. Phys. Lett.* **2014**, *104*, 243105. [CrossRef]

18. Riverola, M.; Sobreviela, G.; Torres, F.; Uranga, A.; Barniol, N. Single-resonator dual frequency BEOL-embedded CMOS-MEMS oscillator with low-power and ultracompact TIA core. *IEEE Electron Device Lett.* **2017**, *38*, 273–276. [CrossRef]

19. Prache, P.; Juillard, J.; Maris Ferreira, P.; Barniol, N.; Riverola, M. Design and characterization of a monolithic CMOS-MEMS mutually injection-locked oscillator for differential resonant sensing. *Sens. Actuators A Phys.* **2018**, *269*, 160–170. [CrossRef]

20. Cai, Z.; van Veldhoven, R.; Suy, H.; de Graaf, G.; Makinwa, K.A.A.; Pertijs, M.A.P. A Phase-Domain Readout Circuit for a CMOS-Compatible Hot-Wire CO_2 Sensor. *IEEE J. Solid State Circuits* **2018**, *53*, 3303–3313. [CrossRef]

21. Uranga, A.; Verd, J.; Barniol, N. CMOS–MEMS resonators: From devices to applications. *Microelectron. Eng.* **2015**, *132*, 58–73. [CrossRef]

22. Ching-Liang, D. A maskless wet etching silicon dioxide post-CMOS process and its application. *Microelectron. Eng.* **2006**, *83*, 2543–2550.

23. Uranga, A.; Sobreviela, G.; Riverola, M.; Torres, F.; Barniol, N. Phase-Noise Reduction in a CMOS-MEMS Oscillator under Nonlinear MEMS Operation. *IEEE Trans. Circuits Syst. I Regul. Pap.* **2017**, *64*, 3047–3055. [CrossRef]

![micromachines logo] *micromachines*

MDPI

Article

A Temperature-Compensated Single-Crystal Silicon-on-Insulator (SOI) MEMS Oscillator with a CMOS Amplifier Chip

Mohammad S. Islam, Ran Wei, Jaesung Lee, Yong Xie, Soumyajit Mandal * and Philip X.-L. Feng *

Department of Electrical Engineering & Computer Science, Case School of Engineering, Case Western Reserve University, Cleveland, OH 44106, USA; msi16@case.edu (M.S.I.); rxw210@case.edu (R.W.); jxl803@case.edu (J.L.); yxx510@case.edu (Y.X.)
* Correspondence: soumyajit.mandal@case.edu (S.M.); philip.feng@case.edu (P.X.-L.F.);
  Tel.: +1-(216)-368-1349 (S.M.); +1-(216)-368-5508 (P.X.-L.F.)

Received: 13 August 2018; Accepted: 18 September 2018; Published: 29 October 2018

Abstract: Self-sustained feedback oscillators referenced to MEMS/NEMS resonators have the potential for a wide range of applications in timing and sensing systems. In this paper, we describe a real-time temperature compensation approach to improving the long-term stability of such MEMS-referenced oscillators. This approach is implemented on a ~26.8 kHz self-sustained MEMS oscillator that integrates the fundamental in-plane mode resonance of a single-crystal silicon-on-insulator (SOI) resonator with a programmable and reconfigurable single-chip CMOS sustaining amplifier. Temperature compensation using a linear equation fit and look-up table (LUT) is used to obtain the near-zero closed-loop temperature coefficient of frequency (TCf) at around room temperature (~25 °C). When subject to small temperature fluctuations in an indoor environment, the temperature-compensated oscillator shows a >2-fold improvement in Allan deviation over the uncompensated counterpart on relatively long time scales (averaging time $\tau > 10,000$ s), as well as overall enhanced stability throughout the averaging time range from $\tau = 1$ to 20,000 s. The proposed temperature compensation algorithm has low computational complexity and memory requirement, making it suitable for implementation on energy-constrained platforms such as Internet of Things (IoT) sensor nodes.

Keywords: oscillator; resonator; micro/nanoelectromechanical systems (MEMS/NEMS); application-specific integrated circuit (ASIC); MEMS-ASIC integration; programmable sustaining amplifier; single-crystal silicon (SC-Si); silicon-on-insulator (SOI); real-time temperature compensation loop

1. Introduction

Stable oscillators are vital for precision timekeeping in various applications, including wired and wireless communications, positioning, navigation, and sensing. In particular, oscillator stability against temperature variations and fluctuations is critical for many of these applications [1,2]. Temperature-compensated quartz crystal oscillators provide excellent stability versus temperature, and have thus dominated the timing and frequency control market for decades. However, they are not suitable for monolithic integration with CMOS circuitry, which makes them sub-optimal for emerging applications such as mobile devices and IoT nodes, where miniaturization is important for reducing cost. Thus, there has been a strong push towards miniaturized alternatives. Oscillators referenced to micromachined resonators [2–5] are promising alternatives due to their small form factors, low phase noise thanks to high Q, low power consumption, good long-term stability, compatibility with batch processing, high reliability, low cost, and wide operating temperature range. It is important to select the right fabrication material in order to optimize such resonators for a particular application. For example, silicon carbide (SiC) has outstanding mechanical or thermal properties, which makes SiC devices suitable for high temperature applications. Here we focus on silicon (Si) devices, since they can be easily integrated with CMOS technology, making them attractive for low-cost applications. Moreover, earlier work has demonstrated the excellent short-term stability of oscillators based on Si MEMS resonators [5]. However, the elastic properties of Si are strongly temperature-dependent, so temperature fluctuations degrade the long-term frequency stability of Si-based oscillators. A number of compensation methods have been reported for improving temperature stability [2,6–13]. For example, in [12], the authors implement a passive compensation method by utilizing silicon dioxide (SiO_2), which has an opposite TCf compared to the structural material of the main resonator (Si); this compensation technique results in a parabolic (second-order) frequency dependence versus temperature. More recently, researchers have proposed the use of multiple temperature compensated resonators with different TCf values to generate a temperature-stable frequency output [13]. These studies, however, have been focused on passive temperature compensation, which increases the complexity of the MEMS design, and therefore, the overall cost of the system.

In this paper, we describe an active real-time software-controlled approach for temperature compensation of MEMS-referenced oscillators (see Figure 1). In particular, we demonstrate our approach using a high-Q single-crystal (SC) Si comb-drive MEMS resonator interfaced with a programmable single-chip CMOS sustaining amplifier. The uncompensated oscillator has a negative TCf, i.e., its oscillation frequency decreases as the temperature increases. In addition, the resonant frequency increases if the DC polarization voltage decreases. We have built a linear regression model based on training data to capture these phenomena, and then used them to develop a real-time temperature compensation loop. The loop is based on (i) simultaneously sampling the oscillator frequency and ambient temperature using a frequency counter and a microcontroller unit (MCU); and (ii) changing the DC polarization voltage to cancel the expected frequency variation. This straightforward technique significantly improves the long-term stability of the oscillator, thus making it suitable for many emerging fundamental and industrial applications [14–20].

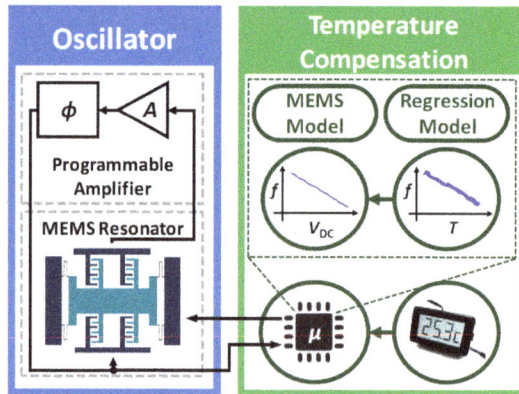

Figure 1. Overview of the temperature-compensated oscillator.

2. System Architecture of the SOI MEMS Resonator and CMOS Sustaining Amplifier

2.1. Overview of the SOI MEMS Resonator

We use SC-Si MEMS comb-drive resonators manufactured in a silicon-on-insulator (SOI) process at wafer scale to ensure high-Q and low phase noise. The SC-Si comb drive devices were fabricated on a 4-inch SOI wafer with a top device layer of 12 µm-thick SC-Si and an oxide (SiO_2) with thickness of 2 µm. We first performed standard photolithography with a resolution of 1 µm to pattern the wafer. The top SC-Si layer not protected by photoresist was etched away by the Bosch process in a deep reactive ion etching (DRIE). The wafer was then diced into 5 mm × 5 mm chips. After etching in buffered oxide etch (BOE) to remove the SiO_2 under the comb drive shuttle, chips were released in a critical point dryer (CPD) to prevent stiction to the substrate. Finally, aluminum (Al) wire bonding was performed on the selective pads for electrical interfacing. As shown in Figure 2a, the fabricated comb-drive SOI MEMS device has an overall length, width, and thickness of 900 μm, 700 μm, and 20 μm, respectively. In addition, the comb-drive fingers are 2 μm apart. We use finite element method (FEM) simulations in COMSOL Multiphysics to study the vibration pattern and mode shape of the first in-plane flexural resonant mode (see Figure 2b). Figure 2 also summarizes the results from optical characterization of the device resonance, which were obtained by using a custom-built laser interferometer specially engineered for ultrasensitive detection of in-plane resonance modes [21,22]. The optically measured resonant frequency for a DC polarization voltage of $V_{DC} = 20$ V is $f \approx 26.8$ kHz at room temperature (the measurement system is shown in Figure S1 in the Supplementary Materials), with a quality factor of $Q = 13,000$ at room temperature and in a moderate vacuum of ~5 mTorr. Figure 2c shows the optothermally excited, optically detected resonance responses [21,22] of the device as a function of temperature (at $V_{DC} = 20$ V); we control the latter with a Peltier cooler. We then extract the resonance frequency versus temperature from the optically measured data (see Figure 2d). The figure shows that this MEMS device has a negative TCf of −34.9 ppm/°C, which is similar to other SC-Si devices in the literature. Figure 2e shows that the resonance frequency also shifts with the DC polarization voltage. We use these empirical measurement results for temperature compensation, as we shall describe in later sections.

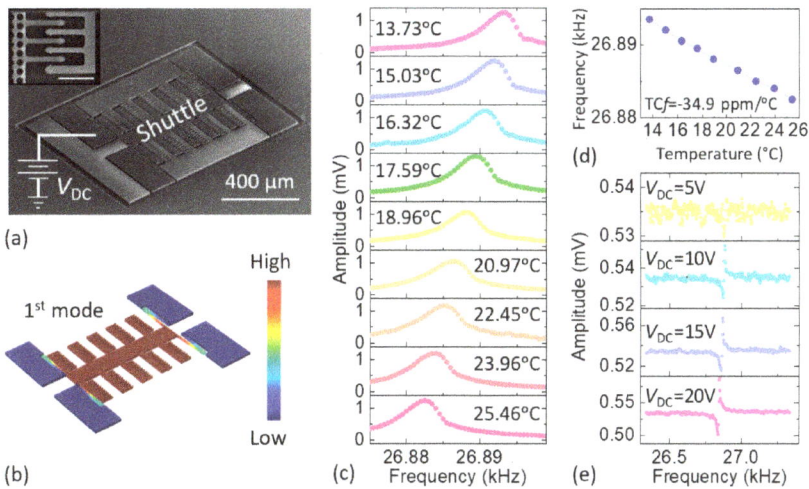

Figure 2. Single-crystal Si-on-insulator (SOI) comb-drive MEMS resonator characteristics. (**a**) Scanning electron microscopy (SEM) image. The inset shows a partial zoom-in view of the comb drive and fingers. Scale bar: 20 μm; (**b**) Simulated vibration pattern and mode shape for the first in-plane resonance mode; (**c**) Optically measured transmission data and frequency response around the first resonance as temperature increases; (**d**) Open-loop resonance frequency dependence on temperature for the extraction of TCf from the data in (**c**); (**e**) Electrically measured transmission and frequency response when various values of the DC polarization voltage are applied to the shuttle. The responses in (**c**,**e**) have not been converted to dimensionless transfer functions since the measurement path includes both electrical and optical components, which makes it difficult to derive absolute calibration factors.

2.2. Overview of the Programmable Single-Chip CMOS Sustaining Amplifier

We have designed and fabricated a single-chip sustaining amplifier using the OnSemi 0.5 μm CMOS process and interfaced it with the MEMS resonator as illustrated in Figure 3a. The chip (see Figure 3b) contains a differential-difference low-noise amplifier (DD-LNA), two second-order band-pass filters (BPFs), three all-pass filters (APFs) used as phase shifters, variable gain amplifiers (VGAs), an automatic level control (ALC), a "background compensation network" (BCN) to cancel the parasitic electrical capacitance of the resonator, and an op-amp-based output buffer [23]. Various parameters of these blocks can be set by using programmable current sources implemented on the test board.

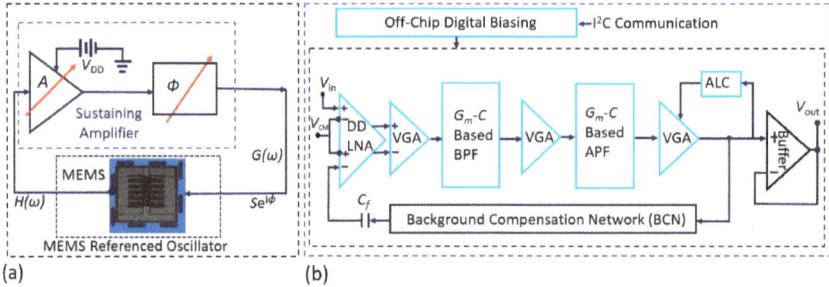

Figure 3. (**a**) Simplified system diagram illustrating the integration of the MEMS resonator chip with the CMOS sustaining amplifier chip; (**b**) Simplified block diagram of the programmable single-chip CMOS sustaining amplifier (corresponding to the same color-coded dashed-line box in (**a**)).

To integrate the SOI MEMS chip and the CMOS amplifier chip, we connect the two differential inputs of the DD-LNA to the MEMS resonator and the BCN, respectively. The DD structure eliminates unwanted capacitive coupling between these terminals. The measured $1/f$ corner frequency of the LNA is <10 kHz. We use a gate-input wide-linear-range operational transconductance amplifier (WLR-OTA) as the basic building block for the rest of the circuit. The WLR-OTA uses source degeneration and bump linearization to improve the input-referred linear range (V_L) [24]. We use this OTA to implement second-order G_m-C BPFs that determine the frequency response of the amplifier, i.e., select a particular resonant mode. Each BPF uses two OTAs in a negative feedback loop to gyrate (invert) the impedance of a capacitor to realize an active inductor, since passive on-chip ones are impractical at such low frequencies. A third OTA acts as a variable resistor to form a parallel RLC circuit, and the OTA bias currents are adjusted to set the resulting center frequency and Q.

Similarly, we use the OTA to implement G_m-C APFs that control the overall phase shift of the amplifier, which enables the phase criterion for oscillations to be satisfied for the chosen mode. Each APF provides unity voltage gain and a frequency-dependent phase shift of $-2\tan^{-1}(\omega t)$, where t = C/G_m. Thus, each APF provides a phase shift of 0–180° as t is adjusted from 0 to ∞ via an off-chip bias currents. However, in practice, t can only be varied over a finite range, which reduces the useful control range to ~120°. Hence three cascaded APFs are used for ~360° control. The maximum signal amplitude for total harmonic distortion (THD) <5% is ≈ 180 mV; this is largely limited by the OTAs.

Four cascaded VGAs control the voltage gain of the amplifier, i.e., enable the gain criterion for oscillations to be satisfied for the chosen mode. Each VGA uses two OTAs to set the gain, and a third OTA to create a high-pass filter. The latter allows the VGAs to be AC-coupled, which prevents accumulation of DC offset and low-frequency interference. The chip also includes an ALC that regulates the oscillator's output voltage amplitude (V_{OUT}) by adjusting the VGA gain. This reduces amplitude-to-phase noise conversion and prevents accidental damage to the MEMS resonator due to overload.

The amplitude- and phase-tunable BCN is designed to cancel the parasitic electrical feedthrough of the MEMS resonator. Such feedthrough, which is generally broadband, makes it difficult for the system to oscillate at the true optimal mechanical resonance frequency [25]. This is undesirable, since off-resonance operation degrades close-in (low offset frequency) phase noise. The BCN drives an on-chip capacitor ($C_f \approx 28$ fF) that feeds back a compensation signal to the negative input terminal of the LNA.

Electrical characterization of the sustaining amplifier shows that the center frequency and Q of the BPF are adjustable from ~10 to ~90 kHz and from 0.5 to 9, respectively, while the APF phase shift can be adjusted over the full 360° range as expected. Moreover, the peak amplifier gain can be adjusted from 0 to 80 dB by using the VGAs, which is sufficient to overcome the transmission loss of typical comb-drive MEMS resonators within this frequency range.

The entire amplifier core occupies a layout area of 1150 μm × 1150 μm, and operates at a supply voltage of 3.3 V. Figure 4a shows a die micrograph of the sustaining amplifier. The input-referred noise of the amplifier with the BPF center frequency set to $f = 26$ kHz is ~7.2 nV/Hz$^{1/2}$, which agrees with circuit simulations and corresponds to an equivalent noise resistance of $R_n = 4$ kΩ at room temperature. Moreover, R_n is typically much smaller than R_m, the motional resistance of comb-drive MEMS resonators in this frequency range. Thus, the close-in phase noise of the oscillator will be dominated by the high-Q resonator, as desired. Details of the measured specifications and performance of the chip are summarized in Table 1.

Table 1. Summary of the electrical specifications and performance of the programmable sustaining amplifier.

Chip Components	Performance
Low-Noise Amplifier (LNA) (for $I_B = 2.5$ μA)	Gain: 12 dB; Bandwidth: ~1 MHz Thermal Noise PSD: 13 nV/Hz$^{1/2}$ $1/f$ Corner Frequency: <10 kHz
Band-Pass Filter (BPF)	Center Frequency (f_0): 2–90 kHz; Q: 1–8 Dynamic Range (DR): 60.7 dB (24 kHz, $Q = 2$) Linear Range: 500 mV (THD < 5%, 24 kHz, $Q = 2$)
All-Pass Filter (APF)	Phase Control: 0–360° Phase Control Sensitivity: ~0.6°/nA
Variable Gain Amplifier (VGA)	Settable Gain: 0–80 dB
Automatic Level Control (ALC)	Amplitude Control Voltage (V_{ED}): 0–0.5 V ED Time Constant: 8-Bit control
Background Compensation Path	Gain Control: −20 to 40 dB Phase Control: 0–180°

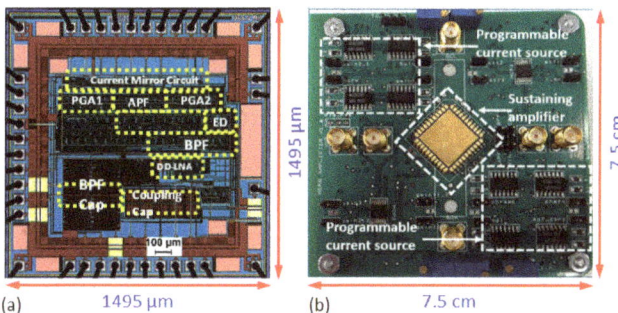

Figure 4. (**a**) Die micrograph of the CMOS sustaining amplifier; (**b**) Test board used for characterizing the amplifier and integrating with the MEMS resonator to build the oscillator.

3. Oscillator Referred to the Single-Crystal SOI MEMS Resonator

3.1. Oscillator System

The packaged sustaining amplifier chip is mounted on a test board designed to fit inside a vacuum chamber (~5 mTorr, room temperature) along with the SOI MEMS resonator. The board only needs four connections: power, output, and a two-wire I²C bus. An external MCU uses the bus to program 16 on-board digital potentiometers. The latter are combined with op-amps to realize programmable bias current generators that, in turn, set all on-chip parameters with 8-bit precision. Figure 5 shows the measured open-loop feed-forward transmission (magnitude and phase) using the MEMS and sustaining amplifier for a DC polarization voltage of V_{DC} = 20 V and a drive amplitude of −10 dBm. We then create a MEMS-referenced oscillator by programming the BPFs, APFs, and VGAs to realize enough gain and the correct phase near resonance. In addition, we program the ALC to set the output amplitude to $V_{OUT} \approx 400$ mV.

The performance of the oscillator is monitored by using a spectrum analyzer (Agilent 4395A, Keysight Technologies, Santa Rosa, CA, USA) and a frequency counter (Agilent 53132A, Keysight Technologies, USA). The measured single-sideband (SSB) phase noise spectrum of the MEMS-referenced oscillator at ambient temperature is shown in Figure 6a, while the dependence of the oscillation frequency on the DC polarization voltage V_{DC} is shown in Figure 6b. Note that the oscillation frequency is a non-monotonic function of V_{DC}.

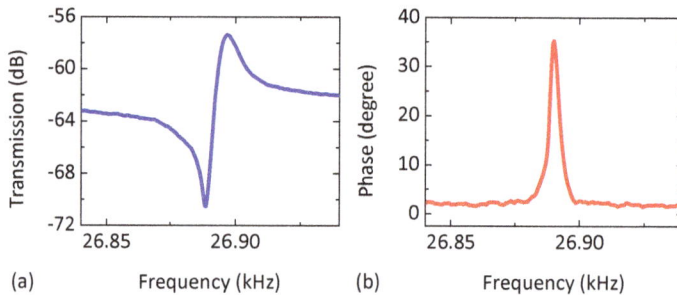

Figure 5. Electrical characterization results of the single-crystal SOI MEMS resonator. (**a**) Measured transmission (dB) and frequency response of the resonance; (**b**) Open-loop phase (degrees) around the first mode for V_{DC} = 20 V and an input power of −10 dBm.

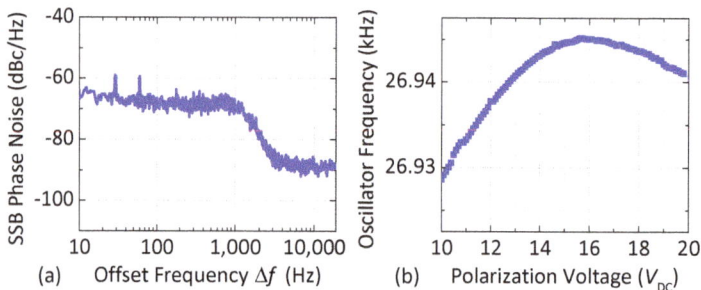

Figure 6. (**a**) Measured single-sideband (SSB) phase noise of the single-crystal SOI MEMS oscillator for V_{DC} = 20 V; (**b**) Measured electrical tuning of the oscillator frequency as a function of V_{DC} (approximately −1 Hz/V for voltages >16 V).

3.2. Temperature Compensation of the MEMS-Referenced Oscillator

Figure 7 shows the block diagram of the overall real-time temperature compensation loop (including calibration of the oscillator using a linear model). The overall real-time temperature compensation loop (i.e., data acquisition, calibration, post-processing, and command execution) has been implemented in a software environment (MATLAB, R2018a). The oscillator is placed inside a vacuum chamber as shown in the left side of the block diagram along with a digital temperature sensor (MCP9808, Microchip Technology, Chandler, AZ, USA) with a resolution of $\pm0.25\ ^\circ$C. The sensor is directly attached to the DIP24 package containing the wire-bonded resonator to ensure that it measures device temperature. A miniaturized development board (Arduino Uno R3, Arduino, New York, NY, USA) based on an ATmega328 MCU (Atmel Corporation, San Jose, CA, USA) with 32 KB memory acts as the system controller. The MCU is used to program 16 on-board digital potentiometers using a 2-wire I^{2}C bus and also acquires data from the temperature sensor. A MATLAB script stores all real-time experimental *T-f* data vectors and sends commands to set the desired MEMS V_{DC} voltage with a resolution of 5 mV using a source meter unit (Keithley 2450, Agilent Technologies, Santa Clara, CA, USA) programmed over GPIB. In addition, the MATLAB script implements temperature-to-frequency (*T-f*) and either frequency-to-voltage (*f-V*) or frequency-to-phase (*f-ϕ*) conversion functions to realize the proposed temperature control loop.

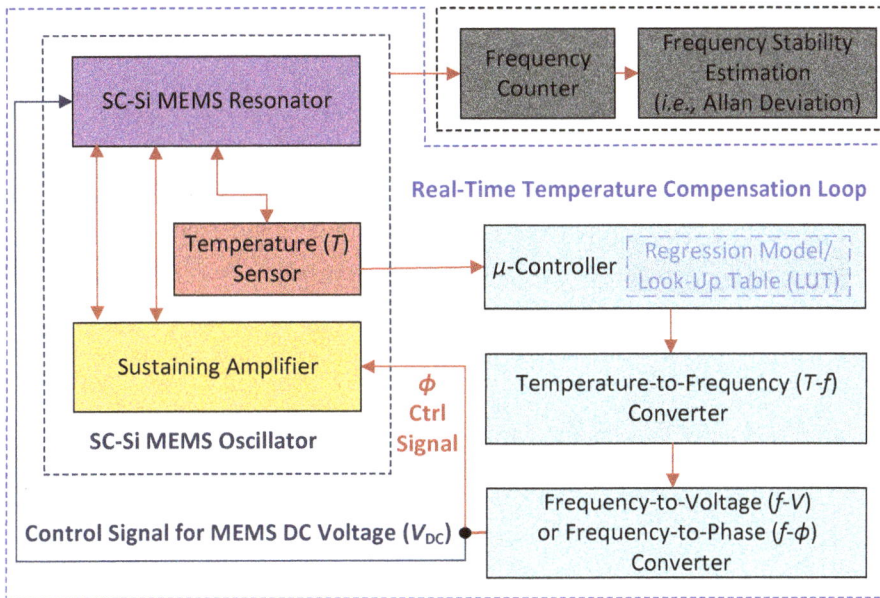

Figure 7. Block diagram of the model calibration procedure, and the proposed real-time temperature compensation loop.

During calibration, the device temperature is accurately set by using a Peltier module (CP08-63-06, Laird Technology, Cleveland, OH, USA) which is powered from an external benchtop DC power supply. The frequency counter has been used to measure the oscillation frequency over a period of seven (7) days with a gate time (t_g) of 100 ms. A data-driven linear model is then derived by using offline post-processing. The best-fitting model given is given by

$$f_{osc} = m \times T + c, \tag{1}$$

where T is the temperature (in °C), f_{osc} is the oscillation frequency (in Hz), m is the slope, and c is the y-intercept. The coefficients of Equation (1) are computed by averaging the best-fitting values from 6 runs (individual values are shown in Table 2). These coefficients are used for temperature compensation. Specifically, Equation (1) is stored in MCU memory and used to calculate the optimal V_{DC} in terms of the measured temperature. Figure 8a shows how the frequency of the uncompensated MEMS-referenced oscillator fluctuates due to room temperature variations; the estimated TCf = −41 ppm/°C is extracted from the fitting in Figure 8b. The result is in good agreement with the open-loop resonator TCf shown in Figure 2d, which suggests that the closed-loop TCf is dominated by the resonator and not the sustaining amplifier. A proportional control algorithm is then used to compensate for the measured TCf, and the resulting frequency stability is estimated using the Allan deviation, $\sigma_A(\tau)$, as a function of averaging time τ [26,27].

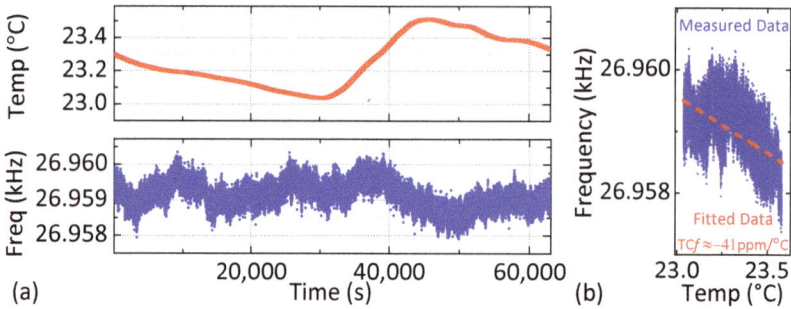

Figure 8. (**a**) Measured temperature variation (top) and fluctuation of oscillation frequency (bottom) for the uncompensated oscillator over 24 h; (**b**) Linear fit of the oscillation frequency versus temperature.

Table 2. Look-up table (LUT) used for temperature compensation.

Day #	Fitted Coefficients	
	m	c
1	−1.022	27,035
2	−1.169	26,964
3	−1.459	27,067
4	−1.128	26,997
5	−1.1329	27,044
6	−1.1458	26,968
Mean value	−1.1761	27,012

Figure 9a,b compare the stability of the temperature-compensated (red curve) and uncompensated (blue curve) oscillators when subject to small indoor temperature fluctuations (approximately ±0.5 °C, detailed statistics are shown in Figure S2 of the Supplementary Materials). The temperature-compensated oscillator achieves a fractional frequency stability of ~3.6 × 10⁻⁶ in the short-term (τ ~1 s), and ~3.7 × 10⁻⁶ for longer averaging times (τ ~1000 s). In the short-term regime (say τ < 100 s), temperature compensation noticeably improves the Allan deviation. Long-term drift dominates at averaging time of τ > 10,000 s; in this region, Allan deviation is significantly improved by compensating for temperature-related drifts in the resonance frequency. We expect such improvements to persist when the proposed temperature compensation method is extended to large batches of wafer-fabricated resonators, since the resonator TCf is mainly determined by the intrinsic material properties of SC-Si.

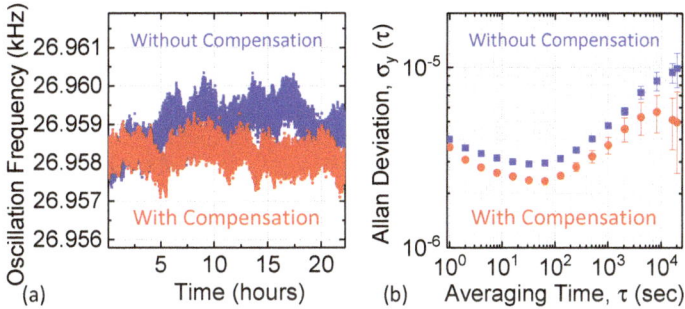

Figure 9. (**a**) Instantaneous frequency of the uncompensated and compensated oscillators over 24 h; (**b**) Measured Allan deviation $\sigma_A (\tau)$ with and without real-time temperature compensation.

We have further verified the performance of the proposed compensation method by measuring oscillator frequency stability over a broader temperature range. For this purpose, we have varied the operating temperature of the resonator from 11.2 °C to 41.2 °C by using either a Peltier module (for cooling) or two ceramic space heaters (for heating). The temperature compensation loop is also modified to control both the MEMS DC polarization voltage (V_{DC}) and the feedback phase shift (ϕ) of the sustaining amplifier to minimize the resulting frequency fluctuations. This is because the oscillation frequency is a non-monotonic function of V_{DC} (see Figure 6b), so it is unsuitable for compensating large temperature variations and drifts. Thus, here we exploit ϕ as another degree of freedom for compensation.

Experimental results are shown in Figure 10 (with supporting data shown in Figure S3 in the Supplementary Materials). The results in Table S1 are plotted in Figure 10. The figure shows that the uncompensated oscillator has the expected TCf of approximately −40 ppm/°C (dominated by the SC-Si resonator), while the compensated oscillator has near-zero TCf (within a measurement error of about ±3 ppm) over this temperature range.

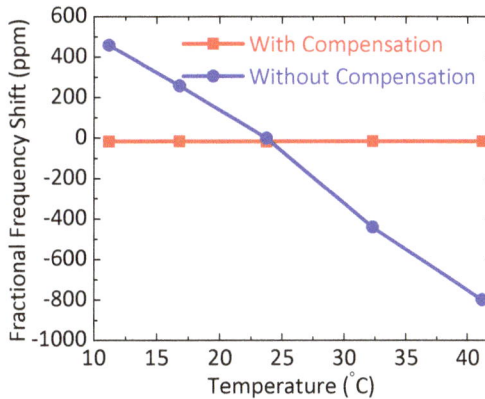

Figure 10. Fractional frequency shift (in ppm) as the device temperature is varied, measured from both the temperature-compensated and uncompensated oscillators over a temperature range of 10 to 45 °C.

Table 3 compares the performance of the proposed temperature-compensated oscillator with other kHz-range MEMS-referenced oscillators. Our design achieves low close-in phase noise and good long-term stability due to temperature compensation.

Table 3. Comparison of performance between the temperature-compensated oscillator in this work and other MEMS-referenced oscillators in the same frequency range.

Properties	This Work	[28]	[29]	[3]
Resonator Type and Material	Single-Crystal SOI Lateral Comb-Drive	3C-SiC, Comb-Drive	Capacitive Transduction H-Shaped Tuning Fork	Poly-Si Two-Port, Folded-Beam, Comb-Drive
Modes of Oscillation	3	3	1	1
Oscillation Frequencies (f_{osc})	~27.0 kHz (Mode 1 Only)	27.1 kHz, 30.3 kHz, 24.2 kHz	32.768 kHz	16.5 kHz
Q-Factor	13,000	13,550,10,300 and 9480	52,000	23,400
CMOS Sustaining Amplifier Chip	0.5 μm CMOS	Discrete Components	180 nm CMOS	CMOS Transimpedance Amplifier (TIA)
Die Size	1.5 mm × 1.5 mm	Discrete Components	1.55 mm × 0.85 mm	420 μm × 320 μm (Resonator Only)
Supply Voltage (V_{DD})	3.3 V	3.0 V	1.4–4.5 V	2.5 V
SSB Phase Noise	−65 dBc/Hz @ 10 Hz Offset	−78 dBc/Hz @ 12 Hz Offset	Not Reported	−72 dBc/Hz @ 1 kHz Offset (Simulated)
FoM *	133.61	Not Reported	Not Reported	Not Reported
Startup Time	~600 μs	Not Reported	0.2 s	Not Reported
Real-Time Temperature Compensation	Yes	No	Yes	No
Temperature coeficients of frequency (TCf)	−34 ppm/°C (MEMS only); <±3 ppm (TCO) over 11.2 °C to +41.2 °C	24.75 ppm /°C ** (rocking vibration mode of MEMS) over 26.85 °C to +426.85 °C	±100 ppm (MEMS only); ±3 ppm (TCXO) & 100 ppm max (XO) over −40 °C to +85 °C	−10 ppm/°C (MEMS only) over 26.85 °C to +96.85 °C
Year	2018	2009	2015	1999

* $FoM = 20 \log\left(\frac{\omega_0}{\Delta\omega}\right) - L(\Delta\omega) - 10 \log\left(\frac{P_c}{1\,\mathrm{mW}}\right)$, ** ANSYS finite element simulation results.

4. Discussion

Temperature-driven frequency fluctuations are ubiquitous, and are a major performance limiter for SC-Si MEMS-referenced oscillators. Other potential sources of frequency fluctuations include noise sources in the instrumentation [30–32], dielectric and charge fluctuations in the resonator, bulk and surface effects in the resonator [33], and limited dynamic range of the sustaining amplifier. The main purpose of the work is to build a temperature-compensated SC-SOI-MEMS oscillator (TCO) for use in low power applications such as IoT sensor nodes. In addition, further improvements in long-term stability can be obtained by replacing temperature compensation with temperature control (i.e., implementing an oven-controlled oscillator (OCO)) and/or by locking the oscillator to an external frequency reference such as GPS over long time scales [34,35]. However, TCOs are more attractive for low-power applications than OCOs because they consume much less power.

5. Conclusions

We have demonstrated improved long-term frequency stability by implementing temperature compensation for a single-crystal SOI MEMS-referenced oscillator with a reconfigurable CMOS sustaining amplifier chip. A software-defined compensation loop has been developed for this purpose. Experimental results confirm the effectiveness of the approach in significantly reducing the Allan deviation on time scales $\tau > 10,000$ s. These results open up new potential applications for

Micromachines **2018**, *9*, 559

MEMS-referenced oscillators, including high-precision sensing, environmental sensing and monitoring, next-generation wireless communication, and navigation.

Supplementary Materials: The following are available online at http://www.mdpi.com/2072-666X/9/11/559/s1, Figure S1. Schematic of the optical measurement system used for characterizing the open-loop resonance response and TCf of the single-crystal SOI MEMS comb-drive resonator. Figure S2. Power spectral density (PSD) and stability estimation: (a) PSD of uncompensated and compensated oscillation frequency; (b) PSD of ambient temperature fluctuations; (c) Allan deviation of uncompensated and compensated oscillation frequency; and (d) Allan deviation of the temperature data shown in (b). Figure S3. Oscillation frequency traces versus time, measured at five different temperature values set and controlled by the sample heating and cooling modules, for oscillators (a) without and (b) with temperature compensation, respectively. Inset plot in (b) shows zoomed-in view of the red box in (b). Table S1. Fractional frequency shift with and without temperature compensation.

Author Contributions: M.S.I. designed the CMOS sustaining amplifier and performed the temperature compensation algorithm and relevant measurements. Y.X. and J.L. designed and fabricated the SOI MEMS resonators. R.W., J.L. and Y.X. measured the open-loop characteristics. M.S.I., R.W. and J.L. performed closed-loop characterization. S.M. and P.X.-L.F. conceived and supervised the project.

Funding: This work was supported by the National Science Foundation (NSF) under grant ECCS-1509721.

Acknowledgments: We would like to thank the MOSIS Educational Program (MEP) for chip fabrication. We are grateful to the financial support from the National Science Foundation (Grant ECCS-1509721).

Conflicts of Interest: The authors declare no conflicts of interest regarding the publication of this paper.

References

1. Vig, J.R. Temperature-insensitive dual-mode resonant sensors—a review. *IEEE Sens. J.* **2001**, *1*, 62–68. [CrossRef]
2. Nguyen, C.T.-C. MEMS technology for timing and frequency control. *IEEE Trans. Ultrason. Ferroelectr. Freq. Control* **2007**, *54*, 251–270. [CrossRef] [PubMed]
3. Nguyen, C.T.-C.; Howe, R.T. An integrated CMOS micromechanical resonator high-Q oscillator. *IEEE J. Solid State Circuits* **1999**, *34*, 440–455. [CrossRef]
4. Thakar, V.; Rais-Zadeh, M. Temperature-compensated piezoelectrically actuated Lame mode resonators. In Proceedings of the IEEE International Conference on Micro Electro Mechanical Systems (MEMS 2014), San Francisco, CA, USA, 26–30 January 2014; pp. 214–217.
5. Lee, H.; Partridge, A.; Assaderaghi, F. Low jitter and temperature stable MEMS oscillators. In Proceedings of the IEEE International Frequency Control Symposium (IFCS), Baltimore, MD, USA, 21–24 May 2012; pp. 266–270.
6. Clark, J.R.; Nguyen, C.T.-C. Mechanically temperature-compensated flexural-mode micromechanical resonators. In Proceedings of the International Electron Devices Meeting, San Francisco, CA, USA, 10–13 December 2000; pp. 399–402.
7. Hsu, W.-T.; Nguyen, C.T.-C. Stiffness-compensated temperature-insensitive micromechanical resonators. In Proceedings of the International Conference on Micro Electro Mechanical Systems (MEMS 2002), Las Vegas, NV, USA, 20–24 January 2002; pp. 731–734.
8. Sundaresan, K.; Allen, P.E.; Ayazi, F. Process and temperature compensation in a 7-MHz CMOS clock oscillator. *IEEE J. Solid State Circuits* **2006**, *41*, 433–442. [CrossRef]
9. Salvia, J.C.; Melamud, R.; Chandorkar, S.A.; Lord, S.F.; Kenny, T.W. Real-time temperature compensation of mems oscillators using an integrated micro-oven and a phase-locked loop. *J. Microelectromech. Syst.* **2010**, *19*, 192–201. [CrossRef]
10. Samarao, A.K.; Ayazi, F. Temperature compensation of silicon resonators via degenerate doping. *IEEE Trans. Electron. Devices* **2012**, *59*, 87–93. [CrossRef]
11. Rais-Zadeh, M.; Thakar, V.A.; Wu, Z.; Peczalski, A. Temperature compensated silicon resonators for space applications. In Proceedings of the Reliability, Packaging, Testing, and Characterization of MOEMS/MEMS and Nanodevices XII, San Francisco, CA, USA, 9 March 2013; SPIE: Bellingham, WA, USA, 2013; Volume 8614.
12. Melamud, R.; Kim, B.; Chandorkar, S.A.; Hopcroft, M.A.; Agarwal, M.; Jha, C.M.; Kenny, T.W. Temperature-compensated high-stability silicon resonators. *Appl. Phys. Lett.* **2007**, *90*, 244107. [CrossRef]

13. Thakar, V.A.; Wu, Z.; Figueroa, C.; Rais-Zadeh, M. A temperature-stable clock using multiple temperature-compensated micro-resonators. In Proceedings of the IEEE International Frequency Control Symposium (IFCS), Taipei, Taiwan, 19–22 May 2014; pp. 1–4.
14. Nguyen, C.T.-C. Micromechanical resonators for oscillators and filters. In Proceedings of the International Ultrasonics Symposium, Seattle, WA, USA, 7–10 November 1995; pp. 489–499.
15. Serrano, D.E.; Tabrizian, R.; Ayazi, F. Tunable piezoelectric MEMS resonators for real-time clock. In Proceedings of the IEEE International Frequency Control Symposium and the European Frequency and Time Forum (IFCS/EFTF), San Francisco, CA, USA, 2–5 May 2011; pp. 1–4.
16. Nguyen, C.T.-C. Frequency-selective MEMS for miniaturized low-power communication devices. *IEEE Trans. Microw. Theory Tech.* **1999**, *47*, 1486–1503. [CrossRef]
17. Perrott, M.H.; Salvia, J.C.; Lee, F.S.; Partridge, A.; Mukherjee, S.; Arft, C.; Jintae, K.; Arumugam, N.; Gupta, P.; Tabatabaei, S.; et al. A temperature-to-digital converter for a MEMS-based programmable oscillator with <±0.5-ppm frequency stability and <1-ps integrated jitter. *IEEE J. Solid State Circuits* **2013**, *48*, 276–291. [CrossRef]
18. Villanueva, L.G.; Kenig, E.; Karabalin, R.B.; Matheny, M.H.; Lifshitz, R.; Cross, M.C.; Roukes, M.L. Surpassing fundamental limits of oscillators using nonlinear resonators. *Phys. Rev. Lett.* **2013**, *110*, 177208. [CrossRef] [PubMed]
19. Villanueva, L.G.; Karabalin, R.B.; Matheny, M.H.; Kenig, E.; Cross, M.C.; Roukes, M.L. A nanoscale parametric feedback oscillator. *Nano Lett.* **2011**, *11*, 5054–5059. [CrossRef] [PubMed]
20. Islam, M.S.; Lee, J.; Wei, R.; Feng, P.X.-L.; Mandal, S.M. Programmable & reconfigurable sustaining amplifiers for MEMS/NEMS referenced multimode oscillators. In Proceedings of the Techical Digest of the 18th Solid-State Sensors, Open Poster WOP-12, Actuators & Microsystems Workshop (Hilton Head 2018), Hilton Head Island, SC, USA, 3–7 June 2018.
21. Lee, J.; Kaul, A.B.; Feng, P.X.-L. Carbon nanofiber high frequency nanomechanical resonators. *Nanoscale* **2017**, *9*, 11864–11870. [CrossRef] [PubMed]
22. McCandless, J.P.; Lee, J.; Kuo, H.I.; Pashaei, V.; Mehregany, M.; Zorman, C.A.; Feng, P.X.-L. Electrical and optical transduction of single-crystal 3C-SiC comb-drive resonators in SiC-on-Insulator (SiCOI) Technology. In Proceedings of the Techical Digest of the 17th Solid-State Sensors, Actuators & Microsystems Workshop (Hilton Head 2016), Hilton Head Island, SC, USA, 5–9 June 2016; pp. 9–12.
23. Islam, M.S.; Singh, S.K.; Mandal, S. A programmable sustaining amplifier for reconfigurable MEMS-referenced oscillators. In Proceedings of the IEEE International Midwest Symposium on Circuits and Systems (MWSCAS), Boston, MA, USA, 6–9 August 2017; pp. 41–44.
24. Sarpeshkar, R.; Lyon, R.F.; Mead, C.A. A low-power wide-linear-range transconductance amplifier. *Analog Integr. Circuits Signal Process.* **1997**, *13*, 123–151. [CrossRef]
25. Feng, X.-L.; White, C.J.; Hajimiri, A.; Roukes, M.L. A self-sustaining ultrahigh-frequency nanoelectromechanical oscillator. *Nat. Nanotechnol.* **2008**, *3*, 342–346. [CrossRef] [PubMed]
26. Allan, D.W. Time and frequency (time-domain) characterization, estimation, and prediction of precision clocks and oscillators. *IEEE Trans. Ultrason. Ferroelectr. Freq. Control* **1987**, *34*, 647–654. [CrossRef] [PubMed]
27. Antonio, D.; Zanette, D.H.; López, D. Frequency stabilization in nonlinear micromechanical oscillators. *Nat. Commun.* **2012**, *3*, 806. [CrossRef] [PubMed]
28. Young, D.J.; Pehlivanoğlu, I.E.; Zorman, C.A. Silicon carbide MEMS-resonator-based oscillator. *J. Micromech. Microeng.* **2009**, *19*, 115027. [CrossRef]
29. Zaliasl, S.; Salvia, J.C.; Hill, G.C.; Chen, L.W.; Joo, K.; Palwai, R.; Arumugam, N.; Phadke, M.; Mukherjee, S.; Lee, H.C.; et al. A 3 ppm 1.5 × 0.8 mm^2 1.0 μA 32.768 kHz MEMS-based oscillator. *IEEE J. Solid State Circuits* **2015**, *50*, 291–302. [CrossRef]
30. Zhang, Y.; Moser, J.; Güttinger, J.; Bachtold, A.; Dykman, M.I. Interplay of driving and frequency noise in the spectra of vibrational systems. *Phys. Rev. Lett.* **2014**, *113*, 255502. [CrossRef] [PubMed]
31. Steele, G.A.; Hüttel, A.K.; Witkamp, B.; Poot, M.; Meerwaldt, H.B.; Kouwenhoven, L.P.; van der Zant, H.S.J. Strong coupling between single-electron tunneling and nanomechanical motion. *Science* **2009**, *325*, 1103–1107. [CrossRef] [PubMed]
32. Miao, T.; Yeom, S.; Wang, P.; Standley, B.; Bockrath, M. Graphene nanoelectromechanical systems as stochastic-frequency oscillators. *Nano Lett.* **2014**, *14*, 2982–2987. [CrossRef] [PubMed]

33. Sansa, M.; Sage, E.; Bullard, E.C.; Gély, M.; Alava, T.; Colinet, E.; Naik, A.K.; Villanueva, L.G.; Duraffourg, L.; Roukes, M.L.; et al. Frequency fluctuations in silicon nanoresonators. *Nat. Nanotechnol.* **2016**, *11*, 552–558. [CrossRef] [PubMed]

34. Cheng, C.L.; Chang, F.R.; Tu, K.Y. Highly accurate real-time GPS carrier phase-disciplined oscillator. *IEEE Trans. Instrum. Meas.* **2005**, *54*, 819–824. [CrossRef]

35. Lombardi, M.A. The use of GPS disciplined oscillators as primary frequency standards for calibration and metrology laboratories. *J. Meas. Sci.* **2008**, *3*, 56–65. [CrossRef]

micromachines

MDPI

Review

Microhotplates for Metal Oxide Semiconductor Gas Sensor Applications—Towards the CMOS-MEMS Monolithic Approach

Haotian Liu [1], Li Zhang [2], King Ho Holden Li [1,2,]* and Ooi Kiang Tan [3]

[1]   School of Mechanical and Aerospace Engineering, Nanyang Technological University, Singapore 639798, Singapore; liuht@ntu.edu.sg
[2]   Temasek Laboratories, Nanyang Technological University, Singapore 67905910, Singapore; li_zhang@ntu.edu.sg
[3]   School of Electrical and Electronic Engineering, Nanyang Technological University, Singapore 67905367, Singapore; eoktan@ntu.edu.sg
*   Correspondence: holdenli@ntu.edu.sg; Tel.: +65-6790-6398

Received: 26 September 2018; Accepted: 26 October 2018; Published: 29 October 2018

Abstract: The recent development of the Internet of Things (IoT) in healthcare and indoor air quality monitoring expands the market for miniaturized gas sensors. Metal oxide gas sensors based on microhotplates fabricated with micro-electro-mechanical system (MEMS) technology dominate the market due to their balance in performance and cost. Integrating sensors with signal conditioning circuits on a single chip can significantly reduce the noise and package size. However, the fabrication process of MEMS sensors must be compatible with the complementary metal oxide semiconductor (CMOS) circuits, which imposes restrictions on the materials and design. In this paper, the sensing mechanism, design and operation of these sensors are reviewed, with focuses on the approaches towards performance improvement and CMOS compatibility.

Keywords: gas sensor; metal oxide (MOX) sensor; micro-electro-mechanical system (MEMS); microhotplate

1. Introduction

Gas sensors have been widely applied in various fields, such as agriculture [1], automotive [2], industrial, indoor air quality monitoring [3] and environmental monitoring [4,5]. Recently, the prevalence of the Internet of Things (IoT) stimulates the development of sensors with small sizes (<10 mm × 10 mm × 10 mm) [6]. In addition, miniaturization of the gas sensors drives the development of electronic noses (E-nose) in various fields, such as food quality control [7,8], disease diagnosis [9,10] and indoor air contaminants classification [11]. Micro-electro-mechanical systems (MEMS) technology is crucial to design and fabricate miniaturized gas sensors with excellent performance such as low power consumption (<100 mW), high sensitivity and fast response/recovery [12]. Additional benefits come from the low cost of the sensor from batch fabrication and the potential to integrate them with signal conditioning circuits.

The gas sensors can be categorized into four types according to Janata [13]: (1) mass sensors; (2) optical sensors; (3) thermal sensors; and (4) electrochemical sensors. A comparison of these four types of sensors is summarized in Table 1, and their relative sizes and power consumptions are shown in Figure 1.

Table 1. Comparison of the four types of gas sensors.

Sensing Principle	Advantages	Disadvantages
Mass	High sensitivity, good reliability, fast response	Piezoelectric substrates are temperature dependent
Optical	High sensitivity, stability over a long lifetime, good selectivity	Difficulty in miniaturization, high cost, high power consumption
Metal oxide (MOX)	Low cost, long lifetime, fast response	Relatively poor selectivity, drift in performance, sensitive to background gas
Thermal (Catalytic)	Low cost, fast response	Detection of flammable gas only, catalyst poisoning, selectivity depends on sensitizers

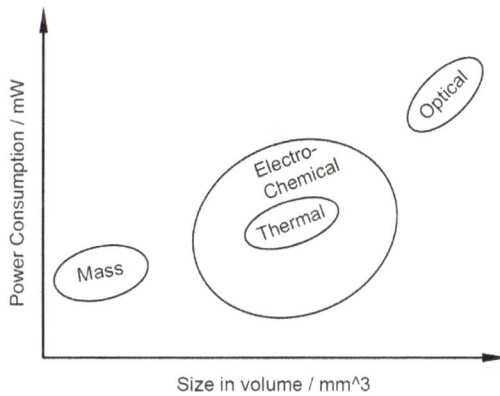

Figure 1. Comparison of the size and power consumption of the four types of gas sensors.

The mass sensors measure the frequency shift of the acoustic wave when foreign particles or molecules are absorbed onto the surface of an oscillating structure. Gravimetric sensors, such as quartz crystal microbalances (QCM), surface acoustic wave (SAW) sensors, surface transverse wave (STW) sensors and shear-horizontal acoustic plate mode (SH-APM) sensors, use quartz crystals with different cuts as the oscillating structure. They are extremely sensitive, with resolutions down to nanogram level. The frequency shift was observed to have an approximately linear relation with the concentration of the gas by Öztürk et al.; additionally, the sensitivity of their QCM sensor can reach a few hertz per ppm [14]. However, the acoustic wave properties heavily depend on the temperature, which gives rise to environment requirements for their applications. Moreover, quartz is not compatible with complementary metal oxide semiconductor (CMOS) processes, and thus, monolithic integration remains challenging.

Thermal sensors detect the gas by the temperature-induced resistance change when combustion reactions take place in catalytic materials. The sensors are low cost and reliable in dusty environments; hence, they are widely used in industrial applications. However, catalytic based reactions restrict the applications of these sensors to flammable gases such as hydrogen, methane, propane and butane. Moreover, catalytic poisoning may deteriorate the long-term stability of the sensors.

Optical sensors recognize the gas by absorption of light with certain frequencies. These frequencies are closely related to the oscillation of molecules; the spectrum suggests that the bonds exist in the molecules. Optical sensors are highly selective and have much higher cost than the other three types because of the complex interferometer mechanisms and the light source. The interferometer and light source also impose challenges on their miniaturization and system integration. Therefore, the high cost and power consumption limit their on-site applications, and most optical sensors still remain as laboratory apparatus [15].

Electrochemical sensors are based on the change in electrical properties when target gas diffuses and reacts with the sensing material. Electrochemical sensors are relatively easy to be miniaturized and manufactured with microfabrication technologies. Successful miniaturizations include field effect transistor (FET) gas sensors and metal oxide semiconductor (MOX). FET sensors are based on the change of the threshold voltage when gas molecules reach the gate material. It is fully compatible with CMOS processes and has a sub-milliwatt level power consumption. However, metal gate materials, such as palladium and platinum, are only sensitive to hydrogen. Recently, many works have been done on the gate design to improve the performance of the FET sensors by applying nanomaterials, such as carbon nanotubes [16] and graphene [17], or by a thin charge inversion layer [18]. MOX gas sensors measure the conductance change of the metal oxide layer and play a dominant role in both research studies and commercial products because they have the most balanced overall performance and low fabrication cost.

Companies such as ams AG [19], Bosch Sensortec [20], Figaro [21] and Sensirion [22] have developed successful commercial MEMS MOX gas sensors for indoor air quality management to detect volatile organic compounds (VOCs). The specifications of some of their products are listed in Table 2. The size of the system depends not only on the hotplate, but also on the package and supporting components. Packaging technology is beyond the scope of this review.

Table 2. Specifications of commercial metal oxide (MOX) gas sensor products.

Company	ams AG			Bosch Sensortec	Figaro	Sensirion
Product No.	iAQ-core P	CCS811	AS-MLV-P2	BME680	TGS8100	SGPC2
Dimension (mm)	15.24×17.78	$2.7 \times 4.0 \times 1.1$	$9.1 \times 9.1 \times 4.5$	$3.0 \times 3.0 \times 0.93$	$3.2 \times 2.5 \times 0.99$	$2.45 \times 2.45 \times 0.9$
Target Gas	CO_2, TVOC	CO_2, TVOC	VOC	TVOC	H_2, C_2H_5OH	TVOC
Power consumption/Current	9 mW	1.2–46 mW	34 mW	<0.1 mA in sleep mode	15 mW	1 mA in low power mode
Supply Voltage (V)	3.3	1.8–3.6	2.7	1.2–3.6	1.8	1.62–1.98
Package	SMD	LGA	-	LGA	Ceramic	DFN
Interface	I²C	I²C	Analog	I²C and SPI	-	I²C

The demand for continuous miniaturization and power consumption reduction drive the research of MOX gas sensors towards direct integration of the MEMS sensing structure with the integrated circuits for signal conditioning circuits [23]. This integration is conventionally realized with the multi-chip approach in which the sensor and circuits are designed and fabricated on separate chips. Multi-chip integration enables independent optimization of the MEMS sensor and CMOS circuits. It also provides more flexibility in the design and fabrication, so that less development time is required. However, extra cost is incurred by the complex packaging and wire bonds. The parasitic element of interconnections gives rise to additional noise. A more advanced way for CMOS-MEMS integration is the monolithic approach, where the sensor and circuits are designed and fabricated on a single substrate. The monolithic integration enhances the performance of the sensor by reducing its size, power consumption and noise [24]. The high development cost and long development cycle can be offset by reduced packaging cost for large-volume manufacturing. The challenge of the monolithic approach is the limitation of materials available for the CMOS processes [25], which will be discussed in detail in this review.

Different aspects of the CMOS-MEMS integration technology have been reviewed by many researchers. For example, Qu focused on the fabrication technologies [26], Li et al. focused on electrochemical biosensors [27], Fischer et al. focused on the integration approaches of MEMS and integrated circuits (IC) and Hierlemann et al. reviewed the fabrication techniques of all the four types of chemical sensors [28]. This review specifically focuses on the recent development in metal oxide CMOS-MEMS gas sensors based on a monolithic approach. The sensing mechanism of metal oxide will be introduced first to define the parameters used to evaluate the performance of the sensor and understand the necessity of the microhotplate in the sensor. The design concerns of the microhotplate are then discussed, with an emphasis on selecting CMOS compatible materials and achieving lower power consumption. Next, the characteristics and synthesis methods of sensing material are briefly discussed. Last, the challenges for integration of the circuit and the sensor are explored.

2. Sensing Mechanism

The reaction between the target gas and the metal oxide film is composed of the reaction of the target gas with metal oxide molecules, as well as the preabsorbed oxygen species. The sensing process consists of three steps: diffusion of the target gas molecules onto the surface of the metal oxide, adsorption of the gas molecules into the metal oxide and reaction between the gas and metal oxide.

In the case of n-type semiconductor metal oxides, the sensing mechanism can be explained by the double Schottky barrier model [29]. The interaction between the metal oxide and the oxygen molecules in air or other oxygen-containing environment generates different oxygen species, such as $O_2(\alpha_1)$, $O_2^-(\alpha_2)$, $O^-(\beta)$ and $O^{2-}(\gamma)$ (Equations (1)–(3)):

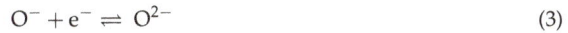

$$O_2 + e^- \rightleftharpoons O_2^- \tag{1}$$

$$O_2^- + e^- \rightleftharpoons 2O^- \tag{2}$$

$$O^- + e^- \rightleftharpoons O^{2-} \tag{3}$$

The released electrons are trapped in the conduction band, forming depletion layers on the surface of the grains. The double Schottky barrier and upward bending of the conduction band at the grain boundaries lead to an increase in the resistance of metal oxide film. When a reducing agent, such as hydrogen, carbon monoxide or ethanol, is brought into contact with the metal oxide film, the resistance drops because of the neutralization of the oxygen species and mitigation of potential barriers. Conversely, an oxidizing agent, such as NO_2, competes with the oxygen species for the electrons and further increases the resistance of the film.

The response of the sensor is defined by the ratio of the resistance of the sensing material before and after exposure to the target gas (Equations (4) and (5)) [29], where R_a and R_g are the resistance of the thin film before and after exposure to the target gas, respectively. Sometimes, the net change in resistance replaces the numerator.

$$S = \frac{R_a}{R_g}, \text{ for n} - \text{type semiconductor} \tag{4}$$

$$S = \frac{R_g}{R_a}, \text{ for p} - \text{type semiconductor} \tag{5}$$

The reaction rate between the chemical species and metal oxide film governs the response of the gas sensor. The α and β species dominate the temperature below and above 300 °C, respectively, while the γ type starts to appear from 550 °C onwards [30]. The reactive species are relatively unstable if the temperature is further increased and the dissolved oxygen atoms compete with the target gas for available active sites. In addition, a further increase in the temperature causes desorption of absorbed gas, causing the sensor response to be further reduced. Therefore, the response of a gas sensor versus the temperature behaves like a bell-shaped curve, with the maximum response obtained at the optimum operating temperature. This optimum temperature varies with respect to the types of metal oxides and the target gas. Therefore, gas sensors usually contain a heating element and an external or internal temperature system to ensure the optimal performance of the sensor.

The response has to be closely related to the centration of gas for effective detection. At a fixed temperature, the sensor response increases with respect to the concentration of the gas due to an increase in carrier concentration and mobility, which can empirically be represented by the power law (Equation (6)) [31]:

$$S = A_g \cdot P_g{}^m \tag{6}$$

where A_g and m are constants and P_g is the partial pressure of the gas, which is proportional to the concentration of the gas. The value of m depends on the concentration of oxygen species when the metal oxide is exposed to air and is specific to each pair of target gas and metal oxide. The value of m

can be approximated by theoretical approaches. Recently, Hua et al. proposed reduced receptor and transducer functions to calculate the value of m [32]. Figure 2a,b show a typical response of WO_3 gas sensor to NO_2 [33]. The effects of concentration on the response of the sensor can be clearly observed.

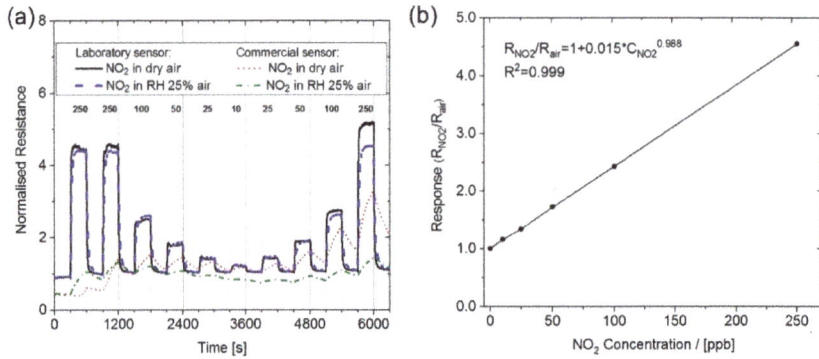

Figure 2. (a) The response of a laboratory WO_3 based gas sensor and commercial MOX sensor with respect to NO_2 in dry and humid air (b) response of the laboratory sensor as a function of NO_2 concentration (from [33]).

Additionally, the sensor response has a strong dependency on the grain size of the metal oxide crystal. Xu et al. [34] found that the resistance of the SnO_2 gas sensor in response to reducing gas would decrease sharply when the grain size fell under two times of the width of the depletion layer, since electron transport was suppressed due to fully depleted grain. Rothschild and Komen showed from a numerical simulation that the sensitivity is proportional to the inverse of the grain size [35]. Yamazoe and Shimanoe proposed the volume and regional depletion concept to formulate the response for small crystals and crystals with planar, spherical and cylindrical shapes [36].

3. MEMS Microhotplate

3.1. Device Layers and Design Considerations

The device structures of a traditional metal oxide gas sensor and a MEMS-based metal oxide gas sensor are shown in Figure 3a,b, respectively. A metal oxide semiconductor gas sensor usually consists of three main components: the microhotplate for temperature elevation, the sensitive material for gas detection and the electrodes for signal transmission. The microhotplate includes a substrate layer, an insulation layer, a heater layer and a passivation layer. The detailed layers are listed as follows:

- The substrate.
- The bottom silicon oxide/nitride layer, which insulates the heating element from the substrate.
- The heating element layer and an adhesion layer, if necessary. Sometimes, temperature sensing elements, such as resistance temperature detectors (RTD), are also fabricated on this layer to monitor the temperature of the microhotplate and provide a reference in the temperature control loop.
- The top silicon nitride/oxide layer, which serves as the insulation between the heater and the sensing material or electrode and passivation layer to prevent catalytic interaction between the target gas and the heater material [37].
- The electrode layer.
- The sensing material layer.

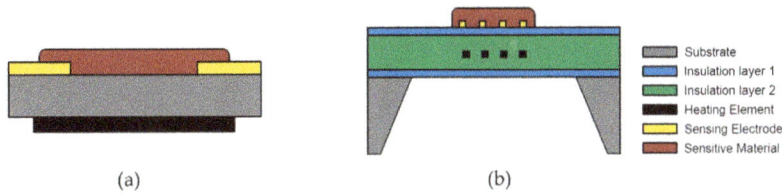

Figure 3. Schematics of (**a**) a traditional metal oxide gas sensor and (**b**) a microhotplate metal oxide gas sensor.

The performance of a sensor is usually evaluated by its 4S, i.e., sensitivity, speed, stability and selectivity. In gas sensor applications, the sensitivity is expressed by the response and detection range. The speed is evaluated by the response and recovery time, which is defined as the time taken to achieve 90% of the change in the resistance. In addition to the 4S, the power consumption of the sensor needs to be minimized due to the voltage and current limit for miniaturized sensors. Extensive studies have been conducted to improve the design of each component for the optimized performance; these will be reviewed accordingly in this section.

The function of the microhotplate is to raise the temperature of the sensitive material to its optimum operating temperature and it is the main source of power consumption in the gas sensor. Bhattacharyya [38] and Spruit [39] have given excellent reviews on the design of microhotplates, concerning the power consumption, temperature homogeneity and mechanical stability.

3.2. Microhotplate

3.2.1. Substrate

Silicon is the mainstream substrate material for micro gas sensors due to its capability with IC fabrication processes and the potential for CMOS integration as a monolithic system for in-situ sensing and processing. Non-silicon materials, such as ceramics (alumina, zirconia), borosilicate glass [40], silicon carbide [41] and plastic [42], are also explored for applications in harsh conditions. These sensors are beyond the scope of this review due to the technological limitations to integrate them with the traditional CMOS circuits at the current stage. The substrate material of all sensors discussed in the following sections of this review is silicon.

3.2.2. Microhotplate Configurations to Reduce Power Consumption

The majority of the power consumed by the microhotplate is converted into thermal energy and dissipated into the surroundings. Therefore, it is essential to suppress the heat loss from the various heat transfer processes, i.e., the conduction from the sensing area to the substrate, the conduction from sensing area to the air, the convection from the sensing area to air and the radiation to the environment. The amount of heat transfer can be obtained from the Equations (7)–(9) [43]:

$$Q_{cond} = -kA_{cond}\nabla T \tag{7}$$

$$Q_{conv} = hA_{conv}(T_s - T_a) \tag{8}$$

$$Q_{rad} = \varepsilon A_{rad}\sigma(T_s^4 - T_b^4) \tag{9}$$

where k is the thermal conductivity of the film material; h is the coefficient of convection of air; ε is the emissivity of the microhotplate; σ is the Stefan-Boltzmann constant; A_{cond}, A_{conv} and A_{rad} are the areas of the surfaces where each heat transfer process takes place; ∇T is the temperature gradient within the solid; T_s is the surface temperature; T_a is the ambient temperature; and T_b is the temperature of the surrounding material, such as the substrate or the package. From Equations (7)–(9), the power consumption can only be reduced by reducing the operating temperature or the area of heat loss. Reduction in the operating temperature can be achieved by depositing metal oxide films with nanostructures, whose response can be compensated by the surface area to volume ratio. Nanostructures and their synthesis method will be discussed in Section 3.4. The latter relies on removing as much material between the heat source and sink as possible and corresponds to the approaches of temperature isolation. Therefore, given the same sensing material, the power consumption can only be minimized by area reduction.

Conduction and convection are the main heat loss mechanisms. Although some studies treated conduction as the primary mechanism [44–46], it is suppressed in microhotplates because of the thin features. Heat loss through convection is comparable and may contribute to up to 60% of the total heat loss in some cases [47,48]. The area in Equation (7) refers to the contact area between the high-temperature zone and the rest of the microhotplate; hence, power consumption can be reduced by geometry optimization. Closed membrane, suspended membrane and bridge (Figure 4a–c, respectively) are the main configurations adopted in previous studies. The substrate right underneath the sensing region is etched away in all three configurations, eliminating the largest contact area. Air acts as heat insulation material due to its much lower thermal conductivity than silicon ($0.018 \, \mathrm{Wm^{-1} \, K^{-1}}$ compared to $145 \, \mathrm{Wm^{-1} \, K^{-1}}$ for silicon).

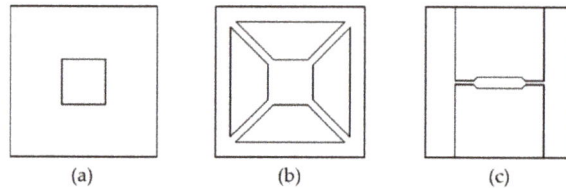

(a)	(b)	(c)

Figure 4. Top view of microhotplates with different configurations: (**a**) closed membrane; (**b**) suspended membrane; (**c**) bridge.

At the expense of reduced power consumption, the mechanical structure of the microhotplate becomes more fragile due to less supporting materials; thus, the configuration must be designed according to the applications. Closed membranes are preferred for commercialized products or on-site applications since the sensors need to withstand vibrations. In research on the other hand, suspended membranes are more popular, since it is able to push the power consumption performance to a limit. In addition, its low thermal inertia contributes to fast response and recovery time and enables reliable temperature modulation. The suspended membrane can either be released from the front side (Figure 5a) or the back side (Figure 5b). Front-side etching requires additional protection of the structure layer or sacrificial layer to create the cavity. Therefore, back-side etching is more common for the sake of simplicity of the processes. Efforts are made to improve the mechanical stability of the suspended structure. Iwata et al. [49] has proposed to reinforce the bridges with the thick polymer layer SU-8 due to its extremely low thermal conductivity ($0.2 \, \mathrm{Wm^{-1}K^{-1}}$) compared to silicon, as well as its compatibility with micromachining processes. However, the reinforced layer increases the fabrication complexity, which is why it has not been adopted in other studies so far. The bridge configuration is an extreme case of the suspended membrane. It further reduces the area for conduction by leaving only two bridges to suspend the hotplate. However, it suffers more from the mechanical instability, as well as the smaller sensing area due to the shrunk hotplate; hence, it is not widely applicable in research works and commercial products.

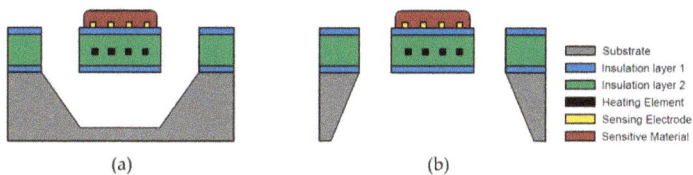

Figure 5. The suspended membrane microhotplate structure formed by (**a**) front-side etching and (**b**) back-side etching.

Both of the areas in Equations (8) and (9) refer to contact area between the microhotplate and air. This area is usually predetermined by the overall size of the device; hence, not much improvement can be made during the design. The convection heat loss to the air can be obtained by fluid simulations [50]. This is ignored in most studies because it only takes up a negligible percentage. Radiation loss soars with the fourth power of the temperature in Kelvin. Pike and Gardner claimed that convection contributes to less than 10% of the total heat loss for applications below 500, according to their lumped element model [51]. Mele et al. claimed from the same model that the radiative loss needs to be taken into account when the temperature reached 600 °C, and is more significant than conductive loss for temperatures above 1000 °C.

A Finite Element Method (FEM) electro-thermal simulation is commonly implemented during the microhotplate design process for optimization of the temperature distribution and power consumption. A large variety of commercially available software such as ANSYS [49,52–56], COMSOL Multiphysics [57–63], ConventorWare [64,65], ISE TCAD [66–68] and MEMCAD [69] has been used. Figure 6 shows the temperature distribution of a suspended membrane microhotplate without a gas sensitive layer for a target temperature of 700 °C [54]. Note that only a quarter of the device was used, as the geometry and symmetry boundary conditions were applied to improve the efficiency of the simulation. In addition to the temperature distribution and uniformity, numerical simulation can also provide critical information for structure design such as the vertical displacement and the maximum stress. However, accurate results are difficult to obtain due to limited access to the properties of thin films and simplifications made in these simulations. The thermal conductivities of the oxide, nitride and metal layers are often directly extracted from open literature. However, the real values vary between different foundries, depending on the conditions of the deposition processes. In addition to the structure layer, introducing the sensing layer makes the thermal analysis of the microhotplate more complicated. The anisotropy of the nanostructures makes its properties hard to be characterized. Moreover, this layer cannot be simply ignored, because the high thermal conductivity of the metal oxide contributes to the temperature uniformity [70]. Therefore, an on-site measurement of thermal conductivity is required; however, no group has done it so far to the best of the author's knowledge.

The temperature profile of the optimized geometry has to be validated by the measurements of an infrared radiation (IR) camera or IR pyrometer. The relationship between the power consumption and the temperature of the hotplate can be obtained by the thermal image together with the measured voltage and current.

The microhotplate is a sandwiched structure formed by the insulation layer, the heater layer and the passivation layer to support the sensing material. Concerns in high-temperature stability determine the material and processes of the sandwiched layers. Residual stress and mismatch in thermal expansion coefficient induce a vertical displacement of the microhotplate at high temperatures and may lead to delamination [48]. Large displacement or weak adhesion might lead to structural failure. Rao et al. reported that the plasma-enhanced chemical vapor deposition (PECVD) nitride layer would peel off easily at a temperature above 691 °C, and thus, the low pressure chemical vapor deposition (LPCVD) nitride layer was adopted instead in their molybdenum hotplate [71]. Low-stress nitride (Si_xN_y) or oxide/nitride/oxide (ONO) layer [63,72–74] is preferred for their low residual stress and good adhesion to metal. Except for nitride and oxide, silicon carbide is also explored as the alternative passivation material by Saxena et al. [75]. A reduced thickness of the material is required to achieve similar thermal behavior, however, the nitride layer is still preferred because of the low cost and simplicity of the CMOS processes.

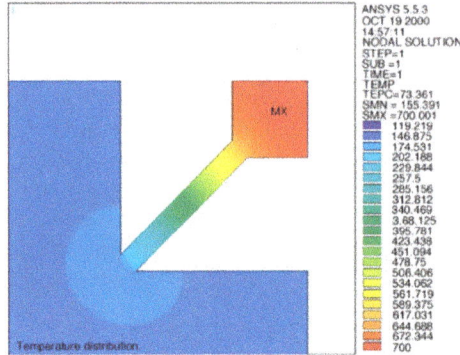

Figure 6. Temperature distribution of a suspended membrane microhotplate without gas sensitive layer (from [54]).

3.2.3. Heater Material

The heating element material has to be stable at high temperature and is preferred to have a linear relationship (Equation (10)) between its resistance and temperature within the operation range:

$$R = R_0 \left(1 + \alpha(T - T_0)\right) \tag{10}$$

where T_0 is the baseline temperature, R_0 is the resistance at the baseline temperature, α is the temperature coefficient of resistance T is the operating temperature and R is the resistance at the operating temperature. The linear behavior assisted the measurement and control of temperature and simplified the signal processing steps. The temperature measurement was either performed by the heater itself or a thermometer on the same plane.

Chemical and mechanical stability at high temperatures were the main concerns for heater material selection. Platinum heaters are the most extensively studied material due to its high thermal conductivity and stability up to 600 °C, which is above the optimum operating temperature of most metal oxides. A layer of titanium or tantalum with tens of nanometers thickness is usually sandwiched between the platinum and insulation layer for better adhesion. For operations above 500 °C, titanium is not recommended, because it will diffuse into the platinum layer and form precipitates on the grain boundaries [76], whereas tantalum shows better behavior due to its function as a diffusion barrier [77,78]. Ceramic adhesion films have also been investigated. Ababneh et al. reported that titanium dioxide adhesion layer would not diffuse into platinum for temperatures up to 800 °C [76]. Halder et al. obtained stable performance of electrodes on a platinized substrate at 1000 °C, with aluminum oxide as the adhesion layer, due to an increase in grain size [79]. Although TiO_2 and Al_2O_3 show much better adhesion performance, their applications are restricted by the feasibility of integration into the CMOS process due to their high-temperature deposition conditions. In addition to single layer adhesion films, a Cr/CrN/Pt/CrN/Cr multilayer approach was demonstrated by Chang and Hsihe [80], which shows improved adhesion and structural stability up to 480 °C. However, the multilayer approach has not been widely applied, because it introduces extra depositing and etching steps to the fabrication processes.

Doped polysilicon is another widely adopted heater material with linear resistance-temperature relations. It is a fully CMOS-compatible material and the adhesion problem no long exists. However, polysilicon is only suitable for operating temperatures below 500 °C, beyond which the recrystallization of polysilicon will cause drift in resistance [81]. Special packaging techniques, such as inert gas sealing, are required to alleviate this problem [82], which are not preferred for commercialization due to its higher cost.

For microhotplates operating above the stability point of platinum and polysilicon, molybdenum [47,71] and tungsten [50,83,84] heaters were studied for applications in harsh conditions. Other than these two metals, Creemer et al. investigated microhotplates based on CMOS compatible titanium nitride (TiN) heater and operating temperature can reach up to 700 °C but the high stress of TiN decrease the yield of the device [48]. The materials above are relatively high cost and mainly used in applications above 300 °C. More affordable materials, such as nickel, serve as better alternatives for operating temperature below 300 °C [55]. Table 3 summarizes the maximum temperature and CMOS compatibility of the various heater materials. Doped polysilicon and tungsten are preferred for monolithic CMOS-MEMS devices due to their compatibility with the processes and their temperature range can cover most operating temperature of the metal oxides.

Table 3. Comparison of maximum temperature and complementary metal oxide semiconductor (CMOS) compatibility among various heater materials.

Heater Material	Maximum Temperature (°C)	CMOS Compatibility	Reference
Pt	600	No	[85]
Doped polysilicon	500	Yes	[81]
Ni	300	No	[55]
Mo	800	No	[71]
W	700	Yes	[86]
TiN	700	Yes	[48]

3.2.4. Heater Geometry to Improve Temperature Uniformity

Both electro- and thermo-migration of the material could cause the degradation of the heater [87]. The former cannot be avoided due to the high operating temperature while the latter depends on the local temperature gradient. Therefore, maintaining the temperature uniformity across the microhotplate is the key to ensure the stable performance of the sensor over its lifetime.

Temperature uniformity can be improved by attaching a layer of material with high thermal conductivity to the microhotplate. The high thermal conductivity material can either be a silicon island [69] left after back side etching or a metal heat spreader plate [49,88] deposited above the heating element (Figure 7).

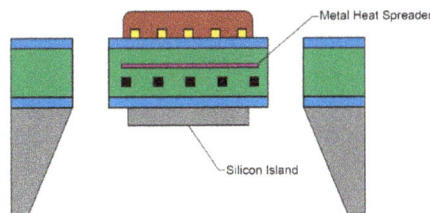

Figure 7. Silicon island and metal heat spreader application on a microhotplate.

Alternatively, optimization of heater geometry with simulation and validation experiments has a significant impact on the temperature uniformity across the microhotplate. Meander, double spiral and drive-wheel are the main geometries reported (Figure 8). Meander shape is, thus far, the most extensively studied one due to its simple geometry. Lee et al. reported that hotspots formation at the center of the meander heater may lead to large temperature variations across the plate [67]. Their numerical simulations and experiments showed remarkably improved temperature uniformity of the drive-wheel type compared to other designs, such as the ultra-low resistance and the honeycomb design.

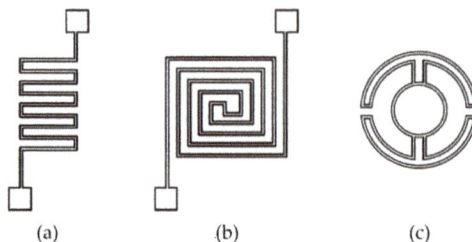

(a) (b) (c)

Figure 8. Heater geometries (**a**) meander, (**b**) double spiral, (**c**) drive wheel.

Except for the sandwiched structure of the microheater, insulation and electrode layers, the heater and electrode can be fabricated on the same plane provided the same material is used (Figure 9a,b). This co-planar design reduces the number of lithography, patterning and etching steps required during the fabrication process. The coplanar heater can also eliminate the parasite capacitance formed between the heater and the sensing layer in sandwiched design [89]. Hwang et al. claimed that coplanar configuration is also advantageous for effective and rapid diffusion.

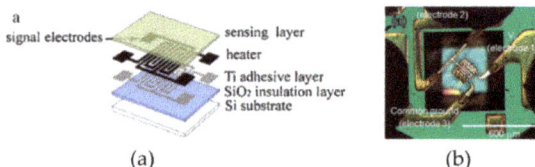

(a) (b)

Figure 9. Coplanar geometry where heating and sensing electrodes lie on the same plane: (**a**) Au-loaded In_2O_3 nanofiber gas sensor by Xu et al. (from [89]) and (**b**) SnO_2 nanowire gas sensor by Hwang et al. (from [90]).

Besides heater electrode deposited on a planar surface, a microhotplate with three-dimensional heater deposited on a three-dimensional V-shape groove was developed for a catalytic type gas sensor [91]. The heating elements lied on a concave surface to obtain a larger active area. However, no metal oxide gas sensor has been developed on the three-dimensional heater thus far, possibly due to the non-uniform temperature distribution across the active area.

3.3. Electrode

The electrode transmits the resistance of the metal oxide film to the signal conditioning circuit during operation: its material is required to have high electrical conductivity, high-temperature stability, good adhesion to the passivation layer and low contact resistance [37]. In addition, the backside/frontside etching to release the membrane is performed after the deposition of the electrode material. Therefore, the electrode material should be inert to the etchant; otherwise, a protection layer is required before the etching process. Most of the electrodes for the MEMS gas sensors are noble metals with an adhesion layer such as Au/Ti, Au/Cr or Pt/Ti. The performance of the electrode will drift after long time operation. Capone et al. observed the degradation of Au/Ti electrodes due to diffusion after the continuous operation at 300 °C and 600 °C. They suggested applying artificial neural network for signal processing to reduce the drift [92]. Noble metal electrodes are believed to have catalytic effects to improve the response of the sensor [93]; however, this point has not been proven yet. In CMOS-MEMS gas sensors, the gold or platinum electrodes are deposited in a post-CMOS process, because they are not CMOS compatible. Sometimes, aluminum is adopted for a full CMOS-compatible process. However, this faces a long-term stability issue under high-temperature operation. The electrode material is deposited in an interdigitated shape so that the large resistance of the metal oxide layer is measurable at low voltage and power consumption. A comprehensive review of the electrode material for both micro gas sensors and traditional gas sensors is given in [94].

3.4. Sensing Material

Eranna has reviewed various types of bulk metal oxide materials and their gas sensing applications [95]. Both n-type (WO_3, SnO_2, ZnO, TiO_2, V_2O_5) and p-type (NiO, CuO, Co_3O_4,) metal oxide semiconductors have been investigated extensively as the sensing materials. However, the n-types takes up more than 90% of the work [96], since the mobility of charge carriers of n-type oxide semiconductors are higher than that of the p-types [30]. Table 4 summarizes the performance of some microhotplate MOX gas sensors.

As the gas sensing reactions occur through gas adsorption, charge transfer and desorption on the surface of the sensing materials, the sensing response is highly dependent on the amount of available surface active sites. In this regard, zero-dimensional nanoparticles with abundant active sites, due to increased specific surface area, are highly desirable for promoting improved gas adsorption and prompting more target gas molecules to participate in the oxidation and reduction reactions. Uniform distribution of nanoparticles with minimum agglomeration is essential to ensure good gas-sensing reactions for this type of material. One-dimensional (1D) nanostructured materials, such as nanowire [84,90,97], nanofiber [89], nanotube [98], nanobelts [99] and nanorods [100] are extensively studied for their high aspect ratio with increased surface-to-volume ratio, giving rise to better stabilities and superior gas-sensing sensitivity with fast response and recovery time. For these 1D nanostructured materials, a smaller grain size/diameter and longer length are, in general, more beneficial for improved gas sensing performance due to the availability of more active sites as the size shrinks and the improved electron transport along the axial direction. Other than these, a lot of research interests have been devoted to two-dimensional (2D) nanostructures for their unique morphologies. These 2D nanostructures, including nanosheets [101] and nanoplates [102], are usually coupled with porous nanostructures and network-like structures to better facilitate the penetration/adsorption of gas molecules into the sensing material, leading to an enhanced sensing response. More recently, three-dimensional (3D) hierarchical nanostructures are receiving increasing attention. These 3D nanostructures with flower-like, leaf-like and spindle-like morphologies are assembled from the low dimensional nanomaterials, including 0D nanoparticles, 1D nanorods and 2D nanosheets [103]. They are deemed to give rise to better sensing performance due to their large surface area, abundant active sites and fast interfacial transport than the low dimensional nanomaterials. Despite of the promising sensing performance, the challenges of inhomogeneity and poor reproducibility have to be addressed for more useful applications of these nanostructures. Noble metals are doped into the metal oxides as chemical or electronic sensitizers to improve the sensitivity, response/recovery time and selectivity of the gas sensor through the chemical or electronic sensitization mechanism [30]. In chemical sensitization, the noble metal assists the adsorption of the target gas by lowering the activation energy. The activated gas is then spilled over the surface of the grain. Most noble metal dopants fall into the electronic sensitization category. The metal particles and their oxidation states act as electrodes attracting the electrons, decreasing the free electrons in the conduction band and creating a depletion region [104]. The electrons return to the grain after exposure to reducing air. This leads to an increase in the resistance and the sensor response (R_a/R_g).

The concentration of noble metal has to be carefully controlled to avoid coagulation. A high percentage of doping may suppress the functionality of the sensor, since agglomeration may decrease the total area of the catalyst surface. Adding catalytic layer results in loss of sensitivity over a long period due to fouling or coking of catalytic metal when exposed to certain gas or vapors. Additionally, the stability of the sensor is affected by the degradation of the dopant metal. The high-temperature degradation of metal oxide materials is not limited by electromigration, but rather by outdiffusion and evaporation of dopants from bulk materials [105].

The metal oxide sensing layer is deposited via a post-processing process after the fabrication of the CMOS circuits and the MEMS microhotplate. Guo et al. have given a comprehensive review of the synthesis methods of metal oxide nanostructures [106]. The sensing material can be synthesized via chemical vapor deposition (CVD), physical vapor deposition (PVD), sol-gel, sintering, spray pyrolysis, spin casting, drop coating, screen printing and ink-jet printing. Among these approaches, CVD, PVD and sol gel are usually employed for thin-film deposition and are compatible with CMOS processes, while the others are more for thick-film with nanostructures. In general, sputtering is preferred over evaporation for its compatibility with a wider choice of materials, better step coverage and enhanced adhesion to the substrate. Surface functionalization or adding binders could further promote the surface adhesion between the sensing materials and substrates [95]. Alternative CMOS compatible ways have been explored. Annanouch et al. grew WO_3/Cu_2O nanoneedles directly on a MEMS hotplate by a single step aerosol-assisted CVD process [107]. The metal oxide nanostructure can also be grown with the internal microheater, whose grain size can be controlled by varying the temperature [108].

3.5. Fabrication Processes

Monolithic gas sensors were fabricated by a CMOS-first approach. The microhotplate materials are deposited and released after the CMOS circuits have been deposited into the wafer. The typical fabrication processes of a suspended membrane microhotplate with a metal heating element are listed as follows and shown in Figure 10:

1. Thermally grown/deposition of the insulation layer.
2. Lift-off/sputtering, patterning and etching of the heater material.
3. Deposition of the passivation layer; patterning and etching for electrical contact.
4. Lift-off/sputtering, patterning and etching of electrode material.
5. Patterning the oxide/nitride layer to define the geometry of the membrane and bridge.
6. Backside/frontside etching to release the suspended membrane.
7. Deposition of the metal oxide layer.

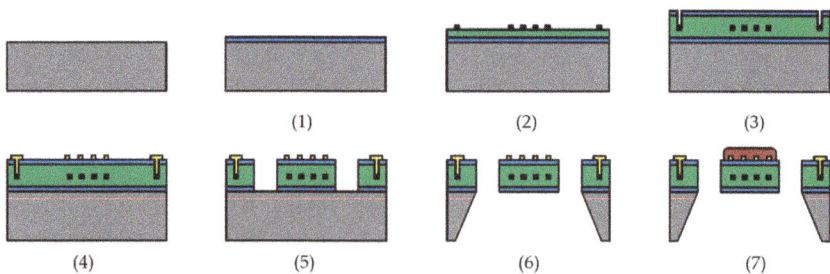

Figure 10. Fabrication processes of a suspended membrane microhotplate with backside etching.

Both dry etching and wet etching have been used in the membrane step. Tetramethylammonium hydroxide (TMAH) or potassium hydroxide (KOH) can be used as the wet etchant, with the former preferred due to full compatibility with CMOS processes. Dry etching processes, such as deep reactive ion etching (DRIE), is, in general, more preferred because the vertical sidewall makes the footprint of the sensor smaller.

Despite the traditional layer-by-layer deposition approach, the silicon on insulator (SOI) CMOS technique was adopted to achieve simple fabrication, the high operating temperature of MOSFETs

(400~600 °C), effective isolation and reduction in leakage current [50,66,88,109]. Either traditional metal heaters [53] or MOSFET heaters [66] can be used for temperature elevation. An example of SOI CMOS microhotplate is shown in Figure 11. The buried oxide layer acts as the etch stop layer during the membrane release step. Additionally, it insulates the heater from the silicon substrate. This technique makes monolithic integration of sensing and signals conditioning units of the system possible.

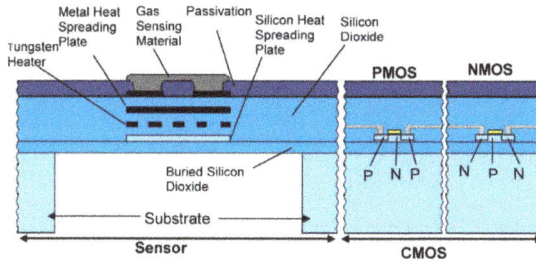

Figure 11. Monolithic gas sensor fabricated with silicon on insulator (SOI) CMOS technology (from [50]).

4. Modes of Operation

The performance of the sensors listed in Table 4 is characterized under the isothermal condition, whereby the temperature and the applied voltage across the heating element remain constant throughout the operation. Low thermal inertia and fast thermal response allow the reduction of power consumption for micromachined gas sensors by pulse mode temperature modulation. In this mode, a square wave voltage is applied across the heater so that the heater is switched on and off. At the expense of reduced power consumption, the interval between two consecutive samplings is longer due to the low response in the low voltage region. For example, the iAQ-core P air quality sensor drops from 66 mW to 9 mW by applying temperature modulation, and the sampling time increases from 1 s to 11 s. Therefore, pulse mode temperature modulation is only suitable for situations when frequent data acquisition is not required. Courbat et al. achieved sub-milliwatt power consumption on their SnO_2/3% Pd CO sensor [110]. Enhancement in the response of the ZnO-nanowire sensor was observed by Shao et al., when it was operating in pulsed mode [84]. They attributed this enhancement to the presence of high-temperature oxygen species during the low-temperature period.

Table 4. Performance of miniaturized metal oxide semiconductor gas sensor devices.

Metal Oxide/Sensitizer	Target Gas	Metal Oxide Morphology	Deposition Process	Heater Material/Geometry	Optimal Temperature (°C)	Power Consumption (mW)	Response	Response/Recovery Time (s)	Detection Limit (ppb)	Reference
CuO	H_2S	Thick film	Drop-cast paste	W/Drive wheel	350	65 @ 600 °C	R_g/R_a = 1.03–1.28 (2–10 ppm)	16.9/49.4	-	[111]
In_2O_3	CH_3CH_2OH	Thin film	Ink jet printing	Pt/Drive wheel	-	24	R_a/R_g = 2.13 (1 ppm)	-	50	[112]
In_2O_3/Au	CH_3CH_2OH	Nanofibers	Electro-spinning + paste	Pt/Meander (Coplanar)	140	222.5	R_a/R_g = 13.8 (500 ppm)	12/14	-	[89]
SnO_2	CH_3CH_2OH	Thin film (various grain sizes)	Chemical vapor deposition (CVD) with metal seed layer	Polysilicon/Meander	400	-	R_a/R_g = 2.1–3.1 (90 ppm)	-	-	[113,114]
SnO_2	CH_3CH_2OH	Nanopore array	Hydro-thermal	Pt/Meander	350	30	R_a/R_g = 1.06 (20 ppb)	1/-	20	[115]
SnO_2	CH_3CH_2OH	Nanowire	Hydro-thermal	Pt/Meander (Coplanar)	500	40	R_a/R_g = 26.2 (100 ppm)	1-2/2.5-3.5	-	[96]
SnO_2	CH_3CH_2OH	Thin film	E-beam evaporation	Pt/Meander	400	9	R_a/R_g = 6.5 (300 ppm)	-	-	[73]
SnO_2	NH_3, CH_3CH_2OH, $(CH_2OH)_2$	Nano-particle	-	Pt/Meander	-	-	R_a/R_g = 6 (100 ppm $(CH_2OH)_2$)	<60/<60	<1000	[116]
SnO_2/Pt, Pd, Au	CH_3CH_2OH, CO, H_2, CH_4	Thin film	Sputtering	Pt/Meander	300	23 mW (annealing @ 950 °C)	R_a/R_g = 2 (100 ppm CH_3CH_2OH)	20-50/10-70	-	[74]
SnO_2/Pd	CH_3CH_2OH	Hollow submicrosphere	Hydro-thermal	Pt/-	300	45	R_a/R_g = -20 (100 ppm)	1.5/18	-	[117]
SnO_2/Pd	H_2	Nanofiber	EHD inkjet printing	W/Meander	185	9.86	R_a/R_g = 8 (2000 ppm)	23/151	-	[118]
SnO_2/Pt	$C_6H_5CH_3$ HCHO	Thin film	RF sputtering	Pt/Meander	300–440	31.5 (HCHO) 45 ($C_6H_5CH_3$)	R_a/R_g = 3.5–4 (HCHO) 3–4 ($C_6H_5CH_3$)	-	-	[119]
SnO_2/Au	CH_4 CO	Thin film	Ion-beam sputtering	Pt/Meander	100 (CO) 250 (CH_4)	20 (CO) 80 (CH_4)	-	-	-	[72]
SnO_2/Sb	CH_3OH	Porous microsphere	LbL deposition + latex removal	Doped polysilicon/Meander	400	-	R_a/R_g = 40.3 (1 ppm)	-	50	[120]
SnO_2-CuO	H_2S	Nanofiber	Electro-spinning	Pt/Meander	200	-	R_a/R_g = 23 (1 ppm)	23/15	<10	[121]
TiO_2	CO	Thin film	Sputtering	Mo/Double spiral	500	104 (@ 800 °C)	R_a/R_g = 6.25 (50 ppm)	0.019/0.034	1000	[71]
TiO_2	CH_3OH	Mesoporous film	Hydro-thermal	Polysilicon/Meander	425	-	R_a/R_g = 7 (50 ppm)	4/30	-	[122]
TiO_2/PdO	H_2	Thin film (180 + 3 nm)	RF sputtering	Pt/Meander	200	48	-	<10/-	-	[63]
WO_3	NO_2	Porous	Drop-casting	W/Drive wheel	300	65 @ 600 °C	R_g/R_a = 5.25 (100 ppb 50% RH air)	40/205	10	[33]

Table 4. *Cont.*

Metal Oxide/Sensitizer	Target Gas	Metal Oxide Morphology	Deposition Process	Heater Material/Geometry	Optimal Temperature (°C)	Power Consumption (mW)	Response	Response/Recovery Time (s)	Detection Limit (ppb)	Reference
WO$_3$/Au-Pt	H$_2$S	Thin film (500 nm)	Sputtering	Pt/Meander	220	-	R$_a$/R$_g$ = 6.5 (1 ppm)	2/30	-	[123]
WO$_3$/Cu$_2$O	H$_2$S	Nanoneedle	Aerosol-assisted CVD	POCl$_3$-doped polysilicon/Double spiral	390	-	R$_a$/R$_g$ = 27.5 (5 ppm)	2/-	<300	[107]
ZnO	H$_2$S	Nanowire	Hydro-thermal	Pt/Drive wheel	300	-	R$_a$/R$_g$ = 1.78 (200 ppb)	-	5	[124]
ZnO	CH$_3$CH$_2$OH	Nanowire	Hydro-thermal	N-doped polysilicon/-	400	33	R$_a$/R$_g$ = 1.6 (809 ppm)	200/~600	-	[97]
ZnO	NH$_3$	Nanowire	Hydro-thermo + Dielectrophoretic	W/Drive wheel	350	55 @ 400 °C	R$_a$/R$_g$ = 4.2 (200 ppm)	228/1290	-	[84]
ZnO/Pd-Ag	CH$_4$	Nano-crystalline	Sol-gel spin coating	Ni/Meander	250	120	R$_a$/R$_g$ = 7.7 (1000 ppm)	8.3/34.8	-	[55]
ZnO-CuO	(CH$_3$)$_2$CO	Nanoflakes	RF sputtering + thermal oxidation	Ni/Double meander (cavity filled)	300	100 @ 259 °C	R$_a$/R$_g$ = 0.46 (10 ppm)	22/26	-	[98]

5. CMOS-MEMS Monolithic Gas Sensor

Although most microhotplates mentioned above are fabricated with CMOS compatible processes, few of them has actual CMOS circuits integrated for temperature control and signal conditioning. The main challenge for CMOS integration is that the operating temperature of the metal oxides (typically 250 to 400 °C) is usually higher than the maximum operating temperature of IC (<300 °C). Therefore, thermal isolation is crucial for monolithic gas sensor design. Suspended membrane geometry is preferred because of their excellent heat isolation performance. Less than 3 °C temperature difference between the circuit and ambient can be achieved by this geometry when the hotplate works at 400 °C [125].

The signal conditioning circuit consists of the temperature control circuit, the resistance readout circuit and the serial bus interface for external communication. In the temperature control circuit, the heater is driven by the MOSFET switches [108] (Figure 12a) with the signal from digital-to-analog converters (DACs). The temperature is controlled by switching the heater on and off alternatively for different time intervals determined by the controller. The controller is of proportional type or proportional-integral-derivative (PID) type to reduce the steady-state error in the temperature control. The actual temperature becomes an input in the feedback loop and is compared with the target temperature. Temperature measurement is done by the on-chip thermometers and the result needs to pass through amplifiers and analog-to-digital converters (ADCs) before processing. The resistance measurement circuit also contains amplifiers and ADCs. In addition, it needs to deal with the large range of the resistance of the metal oxide film from KΩ to MΩ. The resistance signal has to be further processed so that it can be transferred within the allowable number of bits of the converters. The resistance can either be processed by a logarithmic converter [125] (Figure 12b,c) or measured with a biased current [126] (Figure 12d). The recognition of the type of the gas is done by external processors, such as microcontrollers or personal computers. Inter-integrated circuit (I^{2}C) serial protocol is commonly adopted for external communication in both commercialized sensors and the research works because it can reduce the number of pins required.

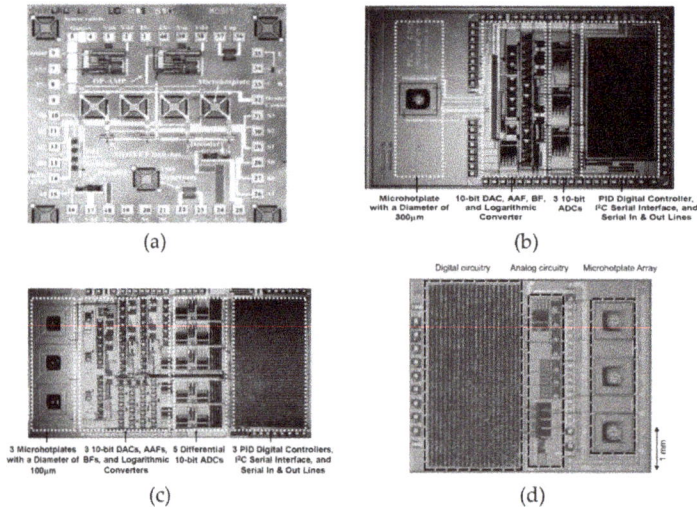

Figure 12. Monolithic CMOS-micro-electro-mechanical systems (MEMS) gas sensors: (**a**) microhotplate array for carbon monoxide detection (from [108]) (**b**) microhotplate gas sensor with proportional-integral-derivative (PID) controllers (from [125]), (**c**) microhotplate gas sensor with mixed-signal architecture (from [125]), (**d**) microhotplate gas sensor using biased current measurement (from [126]).

6. Future Trends and Challenges

It can be observed from Table 2 that most of current commercialized metal oxide gas sensors aim for indoor air quality control and measure the resistance change caused by the adsorption of total volatile organic compounds (TVOC), where the effect of each kind of gas is not distinguished from the others. Recently, learning algorithms such as artificial neural networks have been applied for classification of the volatile organic compounds from the signals generated by e-nose sensor arrays [11]. Moreover, identification of gas has been explored as a technique to diagnose disease [10] and food quality [127]. CMOS-MEMS monolithic gas sensors have great potential for these applications because of its ability to control the operational states of microhotplates coated with different sensing materials. However, the current challenges remain in developing CMOS-compatible approaches to deposit various sensing materials on the microhotplates. Commercial products require uniform performance, and thus, the discrepancies in morphology and thickness should be minimized.

7. Conclusions

The excellent performance of the MEMS metal oxide semiconductor gas sensor makes it dominant in industrial applications and research projects. As the main structure of a gas sensor to operate the sensing material at an elevated temperature, the microhotplate is fabricated into a membrane or bridge for the sake of thermal isolation to achieve low power consumption. Doped polysilicon and tungsten are recommended as the heater material for operating temperatures below 500 °C and 700 °C, respectively, because of their stability in the temperature range and excellent compatibility with the CMOS process. The temperature uniformity can be improved by optimizing the heater geometry through electro-thermal simulations and validation experiments. Besides the power consumption and stability, other performance parameters such as sensitivity, response/recovery time and selectivity are determined by the sensing material deposited with a post processing approach. Recent research in nanostructured materials shows their enhanced gas sensing performance due to the increase in surface area to volume ratio. CMOS-compatible nanomaterial deposition techniques are still being explored.

Monolithic CMOS-MEMS integration has advantages in size, noise level and power consumption compared to the multi-chip integration. However, temperature constraint of the CMOS circuit must always be kept in mind when designing the MEMS process. The CMOS circuits enable precise control of the on/off state of the microhotplates, which is suitable for sensor arrays with different sensing materials on each unit. In addition, the monolithic integration of the MEMS sensor and CMOS circuit make it possible to develop systems in a package (SiP) or system on a chip (SoC) gas sensing in the future. With a microcontroller unit encapsulated in the same package as the sensor, temperature modulation and signal processing become more efficient because of the reduction in noise during signal transmission.

Author Contributions: Conceptualization, H.L., L.Z. and K.H.H.L.; Writing—Original Draft Preparation, H.L.; Writing—Review & Editing, L.Z., K.H.H.L. and O.K.T.; Visualization, H.L.; Supervision, K.H.H.L. and O.K.T.; Funding Acquisition, K.H.H.L.

Funding: Funding support from Temasek Laboratories @ NTU.

Conflicts of Interest: The authors declare no conflict of interest.

References

1. Mitzner, K.D.; Sternhagen, J.; Galipeau, D.W. Development of a micromachined hazardous gas sensor array. *Sens. Actuators B Chem.* **2003**, *93*, 92–99. [CrossRef]
2. Yamazoe, N. Toward innovations of gas sensor technology. *Sens. Actuators B Chem.* **2005**, *108*, 2–14. [CrossRef]
3. Bhattacharya, S.; Sridevi, S.; Pitchiah, R. Indoor Air Quality Monitoring using Wireless Sensor Network. In Proceedings of the 2013 International Conference on Electrical, Electronics and System Engineering (ICEESE), Kuala Lumpur, Malaysia, 4–5 December 2013; pp. 60–64.

4.   Fine, G.F.; Cavanagh, L.M.; Afonja, A.; Binions, R. Metal oxide semi-conductor gas sensors in environmental monitoring. *Sensors* **2010**, *10*, 5469–5502. [CrossRef] [PubMed]

5.   Wetchakun, K.; Samerjai, T.; Tamaekong, N.; Liewhiran, C.; Siriwong, C.; Kruefu, V.; Wisitsoraat, A.; Tuantranont, A.; Phanichphant, S. Semiconducting metal oxides as sensors for environmentally hazardous gases. *Sens. Actuators B Chem.* **2011**, *160*, 580–591. [CrossRef]

6.   Amendola, S.; Lodato, R.; Manzari, S.; Occhiuzzi, C.; Marrocco, G. RFID Technology for IoT-Based Personal Healthcare in Smart Spaces. *IEEE Internet Things J.* **2014**, *1*, 144–152. [CrossRef]

7.   Ghasemi-Varnamkhasti, M.; Mohtasebi, S.S.; Siadat, M.; Balasubramanian, S. Meat Quality Assessment by Electronic Nose (Machine Olfaction Technology). *Sensors* **2009**, *9*, 6058–6083. [CrossRef] [PubMed]

8.   Peris, M.; Escuder-Gilabert, L. A 21st century technique for food control: Electronic noses. *Anal. Chim. Acta* **2009**, *638*, 1–15. [CrossRef] [PubMed]

9.   Gardner, J.W.; Shin, H.W.; Hines, E.L. An electronic nose system to diagnose illness. *Sens. Actuators B Chem.* **2000**, *70*, 19–24. [CrossRef]

10.   Nakhleh, M.K.; Amal, H.; Jeries, R.; Broza, Y.Y.; Aboud, M.; Gharra, A.; Ivgi, H.; Khatib, S.; Badarneh, S.; Har-Shai, L.; et al. Diagnosis and Classification of 17 Diseases from 1404 Subjects via Pattern Analysis of Exhaled Molecules. *ACS Nano* **2017**, *11*, 112–125. [CrossRef] [PubMed]

11.   Zhang, L.; Tian, F.C.; Nie, H.; Dang, L.J.; Li, G.R.; Ye, Q.; Kadri, C. Classification of multiple indoor air contaminants by an electronic nose and a hybrid support vector machine. *Sens. Actuators B Chem.* **2012**, *174*, 114–125. [CrossRef]

12.   Judy, J.W. Microelectromechanical systems (MEMS): Fabrication, design and applications. *Smart Mater. Struct.* **2001**, *10*, 1115–1134. [CrossRef]

13.   Janata, J. *Principles of Chemical Sensors*, 2nd ed.; Springer: Dordrecht, The Netherlands, 2009.

14.   Ozturk, S.; Kosemen, A.; Kosemen, Z.A.; Kilinc, N.; Ozturk, Z.Z.; Penza, M. Electrochemically growth of Pd doped ZnO nanorods on QCM for room temperature VOC sensors. *Sens. Actuators B Chem.* **2016**, *222*, 280–289. [CrossRef]

15.   Hodgkinson, J.; Tatam, R.P. Optical gas sensing: A review. *Meas. Sci. Technol.* **2013**, *24*. [CrossRef]

16.   Zhang, T.; Mubeen, S.; Myung, N.V.; Deshusses, M.A. Recent progress in carbon nanotube-based gas sensors. *Nanotechnology* **2008**, *19*. [CrossRef] [PubMed]

17.   Sharma, B.; Kim, J.S. MEMS based highly sensitive dual FET gas sensor using graphene decorated Pd-Ag alloy nanoparticles for H-2 detection. *Sci. Rep.* **2018**, *8*. [CrossRef] [PubMed]

18.   Fahad, H.M.; Gupta, N.; Han, R.; Desai, S.B.; Javey, A. Highly Sensitive Bulk Silicon Chemical Sensors with Sub-5 nm Thin Charge Inversion Layers. *ACS Nano* **2018**, *12*, 2948–2954. [CrossRef] [PubMed]

19.   Air Quality Sensors. Available online: https://ams.com/air-quality-sensors (accessed on 17 July 2018).

20.   Integrated Environmental Units. Available online: https://www.bosch-sensortec.com/bst/products/environmental/integrated_environmental_unit/overview_integratedenvironmentalunit (accessed on 17 July 2018).

21.   TGS8100. Available online: http://www.figarosensor.com/feature/tgs8100.html (accessed on 17 July 2018).

22.   Multi-Pixel Gas Sensor SGP. Available online: https://www.sensirion.com/en/environmental-sensors/gas-sensors/ (accessed on 17 July 2018).

23.   Gardner, J.W.; Guha, P.K.; Udrea, F.; Covington, J.A. CMOS Interfacing for Integrated Gas Sensors: A Review. *IEEE Sens. J.* **2010**, *10*, 1833–1848. [CrossRef]

24.   Barrettino, D.; Graf, M.; Zimmermann, M.; Hagleitner, C.; Hierlemann, A.; Baltes, H. A smart single-chip micro-hotplate-based gas sensor system in CMOS-technology. *Analog Integr. Circuits Signal Process.* **2004**, *39*, 275–287. [CrossRef]

25.   Hierlemann, A.; Baltes, H. CMOS-based chemical microsensors. *Analyst* **2002**, *128*, 15–28. [CrossRef]

26.   Qu, H. CMOS MEMS Fabrication Technologies and Devices. *Micromachines* **2016**, *7*, 14. [CrossRef]

27.   Li, H.T.; Liu, X.W.; Li, L.; Mu, X.Y.; Genov, R.; Mason, A.J. CMOS Electrochemical Instrumentation for Biosensor Microsystems: A Review. *Sensors* **2017**, *17*. [CrossRef] [PubMed]

28.   Hierlemann, A.; Brand, O.; Hagleitner, C.; Baltes, H. Microfabrication techniques for chemical/biosensors. *Proc. IEEE* **2003**, *91*, 839–863. [CrossRef]

29.   Bârsan, N.; Huebner, M.; Weimar, U. Conduction mechanism in semiconducting metal oxide sensing films: Impact on transduction. In *Semiconductor Gas Sensors*; Elsevier: Amsterdam, The Netherlands, 2013; pp. 35–63.

30. Yamazoe, N.; Sakai, G.; Shimanoe, K. Oxide semiconductor gas sensors. *Catal. Surv. Asia* **2003**, *7*, 63–75. [CrossRef]

31. Yamazoe, N.; Shimanoe, K. Theory of power laws for semiconductor gas sensors. *Sens. Actuators B Chem.* **2008**, *128*, 566–573. [CrossRef]

32. Hua, Z.Q.; Li, Y.; Zeng, Y.; Wu, Y. A theoretical investigation of the power-law response of metal oxide semiconductor gas sensors I: Schottky barrier control. *Sens. Actuators B Chem.* **2018**, *255*, 1911–1919. [CrossRef]

33. Urasinska-Wojcik, B.; Vincent, T.A.; Chowdhury, M.F.; Gardner, J.W. Ultrasensitive WO_3 gas sensors for NO_2 detection in air and low oxygen environment. *Sens. Actuators B Chem.* **2017**, *239*, 1051–1059. [CrossRef]

34. Xu, C.N.; Tamaki, J.; Miura, N.; Yamazoe, N. Grain-Size Effects on Gas Sensitivity of Porous SnO_2-Based Elements. *Sens. Actuators B Chem.* **1991**, *3*, 147–155. [CrossRef]

35. Rothschild, A.; Komem, Y. The effect of grain size on the sensitivity of nanocrystalline metal-oxide gas sensors. *J. Appl. Phys.* **2004**, *95*, 6374–6380. [CrossRef]

36. Yamazoe, N.; Shimanoe, K. New perspectives of gas sensor technology. *Sens. Actuators B Chem.* **2009**, *138*, 100–107. [CrossRef]

37. Korotchenkov, G.S. Handbook of Gas Sensor Materials Properties, Advantages and Shortcomings for Applications Volume 1: Conventional Approaches. In *Integrated analytical systems*; Springer: New York, NY, USA, 2013.

38. Bhattacharyya, P. Technological journey towards reliable microheater development for MEMS gas sensors: A review. *IEEE Trans. Device Mater. Reliab.* **2014**, *14*, 589–599. [CrossRef]

39. Spruit, R.G.; van Omme, J.T.; Ghatkesar, M.K.; Garza, H.H.P. A review on development and optimization of microheaters for high-temperature in situ studies. *J. Microelectromech. Syst.* **2017**, *26*, 1165–1182. [CrossRef]

40. Vasiliev, A.A.; Pisliakov, A.V.; Sokolov, A.V.; Samotaev, N.N.; Soloviev, S.A.; Oblov, K.; Guarnieri, V.; Lorenzelli, L.; Brunelli, J.; Maglione, A.; et al. Non-silicon MEMS platforms for gas sensors. *Sens. Actuators B Chem.* **2016**, *224*, 700–713. [CrossRef]

41. Soo, M.T.; Cheong, K.Y.; Noor, A.F.M. Advances of SiC-based MOS capacitor hydrogen sensors for harsh environment applications. *Sens. Actuators B Chem.* **2010**, *151*, 39–55. [CrossRef]

42. Rieu, M.; Camara, M.; Tournier, G.; Viricelle, J.P.; Pijolat, C.; de Rooij, N.F.; Briand, D. Fully inkjet printed SnO_2 gas sensor on plastic substrate. *Sens. Actuators B Chem.* **2016**, *236*, 1091–1097. [CrossRef]

43. Chen, G. *Nanoscale Energy Transport and Conversion: A Parallel Treatment of Electrons, Phonons, and Photons*; Oxford University Press: New York, NY, USA, 2005; p. xxiii. 531p.

44. Wang, C.Y.; Jin, J.D.; Li, Y.L.; Ding, W.B.; Dai, M.J. Design and fabrication of a MEMS-based gas sensor containing WO3 sensitive layer for detection of NO_2. *J. Micro-Nanolithogr. Mem.* **2017**, *16*. [CrossRef]

45. Bierer, B.; Kneer, J.; Wollenstein, J.; Palzer, S. MEMS based metal oxide sensor for simultaneous measurement of gas induced changes of the heating power and the sensing resistance. *Microsyst. Technol.* **2016**, *22*, 1855–1863. [CrossRef]

46. Saxena, G.; Paily, R. Performance improvement of square microhotplate with insulation layer and heater geometry. *Microsyst. Technol.* **2015**, *21*, 2331–2338. [CrossRef]

47. Mele, L.; Santagata, F.; Iervolino, E.; Mihailovic, M.; Rossi, T.; Tran, A.T.; Schellevis, H.; Creemer, J.F.; Sarro, P.M. A molybdenum MEMS microhotplate for high-temperature operation. *Sens. Actuators A Phys.* **2012**, *188*, 173–180. [CrossRef]

48. Creemer, J.F.; Briand, D.; Zandbergen, H.W.; van der Vlist, W.; de Boer, C.R.; de Rooij, N.F.; Sarro, P.M. Microhotplates with TiN heaters. *Sens. Actuators A Phys.* **2008**, *148*, 416–421. [CrossRef]

49. Iwata, T.; Soo, W.P.C.; Matsuda, K.; Takahashi, K.; Ishida, M.; Sawada, K. Design, fabrication, and characterization of bridgetype micro-hotplates with an SU-8 supporting layer for a smart gas sensing system. *J. Micromech. Microeng.* **2017**, *27*, 024003. [CrossRef]

50. Ali, S.Z.; Udrea, F.; Milne, W.I.; Gardner, J.W. Tungsten-based SOI microhotplates for smart gas sensors. *J. Microelectromech. Syst.* **2008**, *17*, 1408–1417. [CrossRef]

51. Pike, A.; Gardner, J.W. Thermal modelling and characterisation of micropower chemoresistive silicon sensors. *Sens. Actuators B Chem.* **1997**, *45*, 19–26. [CrossRef]

52. Chan, P.C.H.; Yan, G.-Z.; Sheng, L.-Y.; Sharma, R.K.; Tang, Z.; Sin, J.K.O.; Hsing, I.M.; Wang, Y. An integrated gas sensor technology using surface micro-machining. *Sens. Actuators B Chem.* **2002**, *82*, 277–283. [CrossRef]

53. Laconte, J.; Dupont, C.; Flandre, D.; Raskin, J.P. SOI CMOS compatible low-power microheater optimization for the fabrication of smart gas sensors. *IEEE Sens. J.* **2004**, *4*, 670–680. [CrossRef]

54. Belmonte, J.C.; Puigcorbé, J.; Arbiol, J.; Vilà, A.; Morante, J.R.; Sabaté, N.; Gràcia, I.; Cané, C. High-temperature low-power performing micromachined suspended micro-hotplate for gas sensing applications. *Sens. Actuators B Chem.* **2006**, *114*, 826–835. [CrossRef]

55. Bhattacharyya, P.; Basu, P.K.; Mondal, B.; Saha, H. A low power MEMS gas sensor based on nanocrystalline ZnO thin films for sensing methane. *Microelectron. Reliab.* **2008**, *48*, 1772–1779. [CrossRef]

56. Xu, L.; Li, T.; Gao, X.; Wang, Y. Development of a reliable micro-hotplate with low power consumption. *IEEE Sens. J.* **2011**, *11*, 913–919. [CrossRef]

57. Joy, S.; Antony, J.K. In Design and simulation of a micro hotplate using COMSOL multiphysics for MEMS based gas sensor. In Proceedings of the 2015 Fifth International Conference on Advances in Computing and Communications (ICACC), Kochi, India, 2–4 September 2015; pp. 465–468.

58. Elmi, I.; Zampolli, S.; Cozzani, E.; Passini, M.; Cardinali, G.C.; Severi, M. Development of Ultra Low Power Consumption Hotplates for Gas Sensing Applications. *IEEE Sens.* **2006**, *1–3*. [CrossRef]

59. Biró, F.; Dücső, C.; Hajnal, Z.; Riesz, F.; Pap, A.E.; Bársony, I. Thermo-mechanical design and characterization of low dissipation micro-hotplates operated above 500C. *Microelectron. J.* **2014**, *45*, 1822–1828. [CrossRef]

60. Kumar, H.; Singh, K.K.; Sood, N.; Kumar, A.; Mittal, R.K. In Design and simulation of a micro hotplate for MEMS based integrated gas sensing system. In Proceedings of the 2014 IEEE Sensors Applications Symposium (SAS), Queenstown, New Zealand, 18–20 February 2014; pp. 181–184.

61. Noor, M.M.; Sugandi, G.; Aziz, M.F.; Majlis, B.Y. Effects of material and membrane structure on maximum temperature of microheater for gas sensor applications. In Proceedings of the 2014 IEEE International Conference on Semiconductor Electronics (ICSE), Kuala Lumpur, Malaysia, 27–29 August 2014; pp. 255–258.

62. Venkatesh, M.; El Mansouri, B.; Wei, J.; Bossche, A.; Zhang, G. Electro-thermal analysis and design of a combined MEMS impedance and micro hotplate device for gas sensing applications. In Proceedings of the 2016 17th International Conference on Thermal, Mechanical and Multi-Physics Simulation and Experiments in Microelectronics and Microsystems (EuroSimE), Montpellier, France, 17–20 April 2016; pp. 1–9.

63. Kwak, S.; Shim, Y.S.; Yoo, Y.K.; Lee, J.H.; Kim, I.; Kim, J.; Lee, K.H.; Lee, J.H. MEMS-based gas sensor using PdO-decorated TiO$_2$ thin film for highly sensitive and selective H$_2$ detection with low power consumption. *Electron. Mater. Lett.* **2018**, *14*, 305–313. [CrossRef]

64. Tsamis, C.; Nassiopoulou, A.G.; Tserepi, A. Thermal properties of suspended porous silicon micro-hotplates for sensor applications. *Sens. Actuators B Chem.* **2003**, *95*, 78–82. [CrossRef]

65. Prasad, M.; Khanna, V.K. A low-power, micromachined, double spiral hotplate for MEMS gas sensors. *Microsyst. Technol.* **2015**, *21*, 2123–2131. [CrossRef]

66. Udrea, F.; Gardner, J.W.; Setiadi, D.; Covington, J.A.; Dogaru, T.; Lu, C.C.; Milne, W.I. Design and simulations of SOI CMOS micro-hotplate gas sensors. *Sens. Actuators B Chem.* **2001**, *78*, 180–190. [CrossRef]

67. Lee, S.M.; Dyer, D.C.; Gardner, J.W. Design and optimisation of a high-temperature silicon micro-hotplate for nanoporous palladium pellistors. *Microelectron. J.* **2003**, *34*, 115–126. [CrossRef]

68. Lu, C.-C.; Liao, K.-H.; Udrea, F.; Covington, J.A.; Gardner, J.W. Multi-field simulations and characterization of CMOS-MEMS high-temperature smart gas sensors based on SOI technology. *J. Micromech. Microeng.* **2008**, *18*, 075010. [CrossRef]

69. Briand, D.; Heimgartner, S.; Gretillat, M.A.; van der Schoot, B.; de Rooij, N.F. Thermal optimization of micro-hotplates that have a silicon island. *J. Micromech. Microeng.* **2002**, *12*, 971–978. [CrossRef]

70. Graf, M.; Jurischka, R.; Barrettino, D.; Hierlemann, A. 3D nonlinear modeling of microhotplates in CMOS technology for use as metal-oxide-based gas sensors. *J. Micromech. Microeng.* **2005**, *15*, 190–200. [CrossRef]

71. Rao, L.L.R.; Singha, M.K.; Subramaniam, K.M.; Jampana, N.; Asokan, S. Molybdenum microheaters for MEMS-based gas sensor applications: Fabrication, electro-thermo-mechanical and response characterization. *IEEE Sens. J.* **2017**, *17*, 22–29.

72. Kim, I.; Seo, K.W.; Kim, I. Ultra-thin filmed SnO$_2$ gas sensor with a lowpower micromachined hotplate for selective dual gas detection of carbon monoxide and methane. In Proceedings of the 2017 Eleventh International Conference on Sensing Technology (ICST), Sydney, Australia, 4–6 December 2017; pp. 1–5.

73. Mo, Y.; Okawa, Y.; Tajima, M.; Nakai, T.; Yoshiike, N.; Natukawa, K. Micro-machined gas sensor array based on metal film micro-heater. *Sens. Actuators B Chem.* **2001**, *79*, 175–181. [CrossRef]

74. Bin, G.; Bermak, A.; Chan, P.C.H.; Yan, G.Z. An Integrated Surface Micromachined Convex Microhotplate Structure for Tin Oxide Gas Sensor Array. *IEEE Sens. J.* **2007**, *7*, 1720–1726.

75. Saxena, G.; Paily, R. Choice of insulation materials and its effect on the performance of square microhotplate. *Microsyst. Technol.* **2015**, *21*, 393–399. [CrossRef]

76. Ababneh, A.; Al-Omari, A.N.; Dagamseh, A.M.K.; Tantawi, M.; Pauly, C.; Mucklich, F.; Feili, D.; Seidel, H. Electrical and morphological characterization of platinum thin-films with various adhesion layers for high temperature applications. *Microsyst. Technol.* **2017**, *23*, 703–709. [CrossRef]

77. Tiggelaar, R.M.; Sanders, R.G.R.; Groenland, A.W.; Gardeniers, J.G.E. Stability of thin platinum films implemented in high-temperature microdevices. *Sens. Actuators A Phys.* **2009**, *152*, 39–47. [CrossRef]

78. Marasso, S.L.; Tommasi, A.; Perrone, D.; Cocuzza, M.; Mosca, R.; Villani, M.; Zappettini, A.; Calestani, D. A new method to integrate ZnO nano-tetrapods on MEMS micro-hotplates for large scale gas sensor production. *Nanotechnology* **2016**, *27*. [CrossRef] [PubMed]

79. Halder, S.; Schneller, T.; Waser, R. Enhanced stability of platinized silicon substrates using an unconventional adhesion layer deposited by CSD for high temperature dielectric thin film deposition. *Appl. Phys. A* **2007**, *87*, 705–708. [CrossRef]

80. Chang, W.-Y.; Hsihe, Y.-S. Multilayer microheater based on glass substrate using MEMS technology. *Microelectron. Eng.* **2016**, *149*, 25–30. [CrossRef]

81. Ehmann, M.; Ruther, P.; von Arx, M.; Paul, O. Operation and short-term drift of polysilicon-heated CMOS microstructures at temperatures up to 1200 K. *J. Micromech. Microeng.* **2001**, *11*, 397–401. [CrossRef]

82. Samaeifar, F.; Hajghassem, H.; Afifi, A.; Abdollahi, H. Implementation of high-performance MEMS platinum micro-hotplate. *Sens. Rev.* **2015**, *35*, 116–124. [CrossRef]

83. Wang, J.; Yu, J. Multifunctional platform with CMOS-compatible tungsten microhotplate for pirani, temperature, and gas sensor. *Micromachines* **2015**, *6*, 1597–1605. [CrossRef]

84. Shao, F.; Fan, J.D.; Hernandez-Ramirez, F.; Fabrega, C.; Andreu, T.; Cabot, A.; Prades, J.D.; Lopez, N.; Udrea, F.; De Luca, A.; et al. NH₃ sensing with self-assembled ZnO-nanowire μHP sensors in isothermal and temperature-pulsed mode. *Sens. Actuators B Chem.* **2016**, *226*, 110–117. [CrossRef]

85. Firebaugh, S.L.; Jensen, K.F.; Schmidt, M.A. Investigation of high-temperature degradation of platinum thin films with an in situ resistance measurement apparatus. *J. Microelectromech. Syst.* **1998**, *7*, 128–135. [CrossRef]

86. Haque, M.S.; Teo, K.B.K.; Rupensinghe, N.L.; Ali, S.Z.; Haneef, I.; Maeng, S.; Park, J.; Udrea, F.; Milne, A.I. On-chip deposition of carbon nanotubes using CMOS microhotplates. *Nanotechnology* **2008**, *19*. [CrossRef] [PubMed]

87. Biro, F.; Hajnal, Z.; Dusco, C.; Barsony, I. The critical impact of temperature gradients on Pt filament failure. *Microelectron. Reliab.* **2017**, *78*, 118–125. [CrossRef]

88. Guha, P.K.; Ali, S.Z.; Lee, C.C.C.; Udrea, F.; Milne, W.; Iwaki, T.; Covington, J.; Gardner, J. Novel design and characterisation of SOI CMOS micro-hotplates for high temperature gas sensors. *Sens. Actuators B Chem.* **2007**, *127*, 260–266. [CrossRef]

89. Xu, X.J.; Fan, H.T.; Liu, Y.T.; Wang, L.J.; Zhang, T. Au-loaded In₂O₃ nanofibers-based ethanol micro gas sensor with low power consumption. *Sens. Actuators B Chem.* **2011**, *160*, 713–719. [CrossRef]

90. Hwang, I.S.; Lee, E.B.; Kim, S.J.; Choi, J.K.; Cha, J.H.; Lee, H.J.; Ju, B.K.; Lee, J.H. Gas sensing properties of SnO₂ nanowires on micro-heater. *Sens. Actuators B Chem.* **2011**, *154*, 295–300. [CrossRef]

91. Xu, L.; Li, T.; Gao, X.L.; Wang, Y.L. A High-Performance Three-Dimensional Microheater-Based Catalytic Gas Sensor. *IEEE Electr. Device Lett.* **2012**, *33*, 284–286. [CrossRef]

92. Capone, S.; Epifani, M.; Francioso, L.; Kaciulis, S.; Mezzi, A.; Siciliano, P.; Taurino, A.M. Influence of electrodes ageing on the properties of the gas sensors based on SnO₂. *Sens. Actuators B Chem.* **2006**, *115*, 396–402. [CrossRef]

93. Bârsan, N.; Weimar, U. Conduction Model of Metal Oxide Gas Sensors. *J. Electroceram.* **2001**, *7*, 143–167. [CrossRef]

94. Lee, S.P. Electrodes for Semiconductor Gas Sensors. *Sensors* **2017**, *17*, 683. [CrossRef] [PubMed]

95. Eranna, G.; Joshi, B.C.; Runthala, D.P.; Gupta, R.P. Oxide materials for development of integrated gas sensors—A comprehensive review. *Crit. Rev. Solid State Mater. Sci.* **2004**, *29*, 111–188. [CrossRef]

96. Kim, H.J.; Lee, J.H. Highly sensitive and selective gas sensors using p-type oxide semiconductors: Overview. *Sens. Actuators B Chem.* **2014**, *192*, 607–627. [CrossRef]

97. Santra, S.; Guha, P.K.; Ali, S.Z.; Hiralal, P.; Unalan, H.E.; Covington, J.A.; Amaratunga, G.A.J.; Milne, W.I.; Gardner, J.W.; Udrea, F. ZnO nanowires grown on SOI CMOS substrate for ethanol sensing. *Sens. Actuators B Chem.* **2010**, *146*, 559–565. [CrossRef]

98. Behera, B.; Chandra, S. An innovative gas sensor incorporating ZnO–CuO nanoflakes in planar MEMS technology. *Sens. Actuators B Chem.* **2016**, *229*, 414–424. [CrossRef]

99. Sadek, A.Z.; Choopun, S.; Wlodarski, W.; Ippolito, S.J.; Kalantar-zadeh, K. Characterization of ZnO nanobelt-based gas sensor for H_2, NO_2, and hydrocarbon sensing. *IEEE Sens. J.* **2007**, *7*, 919–924. [CrossRef]

100. Wang, L.W.; Kang, Y.F.; Liu, X.H.; Zhang, S.M.; Huang, W.P.; Wang, S.R. ZnO nanorod gas sensor for ethanol detection. *Sens. Actuators B Chem.* **2012**, *162*, 237–243. [CrossRef]

101. Li, Z.J.; Lin, Z.J.; Wang, N.N.; Wang, J.Q.; Liu, W.; Sun, K.; Fu, Y.Q.; Wang, Z.G. High precision NH_3 sensing using network nano-sheet Co_3O_4 arrays based sensor at room temperature. *Sens. Actuators B Chem.* **2016**, *235*, 222–231. [CrossRef]

102. Shendage, S.S.; Patil, V.L.; Vanalakar, S.A.; Patil, S.P.; Harale, N.S.; Bhosale, J.L.; Kim, J.H.; Patil, P.S. Sensitive and selective NO_2 gas sensor based on WO3 nanoplates. *Sens. Actuators B Chem.* **2017**, *240*, 426–433. [CrossRef]

103. Lin, T.T.; Lv, X.; Li, S.; Wang, Q.J. The Morphologies of the Semiconductor Oxides and Their Gas-Sensing Properties. *Sensors* **2017**, *17*. [CrossRef] [PubMed]

104. Fu, J.C.; Zhao, C.H.; Zhang, J.L.; Peng, Y.; Xie, E.Q. Enhanced Gas Sensing Performance of Electrospun Pt-Functionalized NiO Nanotubes with Chemical and Electronic Sensitization. *ACS Appl. Mater. Interfaces* **2013**, *5*, 7410–7416. [CrossRef] [PubMed]

105. Spannhake, J.; Helwig, A.; Müller, G.; Faglia, G.; Sberveglieri, G.; Doll, T.; Wassner, T.; Eickhoff, M. SnO_2:Sb–A new material for high-temperature mems heater applications: Performance and limitations. *Sens. Actuators B Chem.* **2007**, *124*, 421–428. [CrossRef]

106. Guo, T.; Yao, M.S.; Lin, Y.H.; Nan, C.W. A comprehensive review on synthesis methods for transition-metal oxide nanostructures. *CrystEngComm* **2015**, *17*, 3551–3585. [CrossRef]

107. Annanouch, F.E.; Haddi, Z.; Vallejos, S.; Umek, P.; Guttmann, P.; Bittencourt, C.; Llobet, E. Aerosol-Assisted CVD-Grown WO_3 Nanoneedles Decorated with Copper Oxide Nanoparticles for the Selective and Humidity-Resilient Detection of H_2S. *ACS Appl. Mater. Interfaces* **2015**, *7*, 6842–6851. [CrossRef] [PubMed]

108. Afridi, M.Y.; Suehle, J.S.; Zaghloul, M.E.; Berning, D.W.; Hefner, A.R.; Cavicchi, R.E.; Semancik, S.; Montgomery, C.B.; Taylor, C.J. A monolithic CMOS microhotplate-based gas sensor system. *IEEE Sens. J.* **2002**, *2*, 644–655. [CrossRef]

109. Graf, M.; Gurlo, A.; Barsan, N.; Weimar, U.; Hierlemann, A. Microfabricated gas sensor systems with sensitive nanocrystalline metal-oxide films. *J. Nanoparticle Res.* **2006**, *8*, 823–839. [CrossRef]

110. Courbat, J.; Briand, D.; Yue, L.; Raible, S.; de Rooij, N.F. Drop-coated metal-oxide gas sensor on polyimide foil with reduced power consumption for wireless applications. *Sens. Actuators B Chem.* **2012**, *161*, 862–868. [CrossRef]

111. Urasinska-Wojcik, B.; Gardner, J.W. H_2S sensing in dry and humid H_2 environment with p-type CuO thick-film gas sensors. *IEEE Sens. J.* **2018**, *18*, 3502–3508. [CrossRef]

112. Moon, S.E.; Lee, H.K.; Choi, N.J.; Kang, H.T.; Lee, J.; Ahn, S.D.; Kang, S.Y. Low power consumption micro C_2H_5OH gas sensor based on micro-heater and ink jetting technique. *Sens. Actuators B Chem.* **2015**, *217*, 146–150. [CrossRef]

113. Panchapakesan, B.; DeVoe, D.L.; Widmaier, M.R.; Cavicchi, R.; Semancik, S. Nanoparticle engineering and control of tin oxide microstructures for chemical microsensor applications. *Nanotechnology* **2001**, *12*, 336–349. [CrossRef]

114. Semancik, S.; Cavicchi, R.E.; Wheeler, M.C.; Tiffany, J.E.; Poirier, G.E.; Walton, R.M.; Suehle, J.S.; Panchapakesan, B.; DeVoe, D.L. Microhotplate platforms for chemical sensor research. *Sens. Actuators B Chem.* **2001**, *77*, 579–591. [CrossRef]

115. Xu, L.; Dai, Z.; Duan, G.; Guo, L.; Wang, Y.; Zhou, H.; Liu, Y.; Cai, W.; Wang, Y.; Li, T. Micro/nano gas sensors: A new strategy towards in-situ wafer-level fabrication of high performance gas sensing chips. *Sci. Rep.* **2015**, *5*. [CrossRef] [PubMed]

116. Wang, J.; Yang, J.; Chen, D.; Jin, L.; Li, Y.; Zhang, Y.; Xu, L.; Guo, Y.; Lin, F.; Wu, F. Gas detection microsystem with MEMS gas sensor and integrated circuit. *IEEE Sens. J.* **2018**. [CrossRef]

117. Xiao, L.; Xu, S.R.; Yu, G.; Liu, S.T. Efficient hierarchical mixed Pd/SnO$_2$ porous architecture deposited microheater for low power ethanol gas sensor. *Sens. Actuators B Chem.* **2018**, *255*, 2002–2010. [CrossRef]

118. Wu, H.; Yu, J.; Cao, R.; Yang, Y.; Tang, Z. Electrohydrodynamic inkjet printing of Pd loaded SnO$_2$ nanofibers on a CMOS micro hotplate for low power H$_2$ detection. *AIP Adv.* **2018**, *8*, 055307. [CrossRef]

119. Kang, J.G.; Park, J.S.; Lee, H.J. Pt-doped SnO$_2$ thin film based micro gas sensors with high selectivity to toluene and HCHO. *Sens. Actuators B Chem.* **2017**, *248*, 1011–1016. [CrossRef]

120. Martinez, C.J.; Hockey, B.; Montgomery, C.B.; Semancik, S. Porous tin oxide nanostructured microspheres for sensor applications. *Langmuir* **2005**, *21*, 7937–7944. [CrossRef] [PubMed]

121. Zhao, Y.; He, X.L.; Li, J.P.; Gao, X.G.; Jia, J. Porous CuO/SnO$_2$ composite nanofibers fabricated by electrospinning and their H$_2$S sensing properties. *Sens. Actuators B Chem.* **2012**, *165*, 82–87. [CrossRef]

122. Benkstein, K.D.; Semancik, S. Mesoporous nanoparticle TiO$_2$ thin films for conductometric gas sensing on microhotplate platforms. *Sens. Actuators B Chem.* **2006**, *113*, 445–453. [CrossRef]

123. Tao, W.H.; Tsai, C.H. H2S sensing properties of noble metal doped WO$_3$ thin film sensor fabricated by micromachining. *Sens. Actuators B Chem.* **2002**, *81*, 237–247. [CrossRef]

124. Chen, Y.; Xu, P.C.; Xu, T.; Zheng, D.; Li, X.X. ZnO-nanowire size effect induced ultra-high sensing response to ppb-level H$_2$S. *Sens. Actuators B Chem.* **2017**, *240*, 264–272. [CrossRef]

125. Barrettino, D.; Graf, M.; Taschini, S.; Hafizovic, S.; Hagleitner, C.; Hierlemann, A. CMOS monolithic metal–oxide gas sensor microsystems. *IEEE Sens. J.* **2006**, *6*, 276–286. [CrossRef]

126. Frey, U.; Graf, M.; Taschini, S.; Kirstein, K.U.; Hierlemann, A. A digital CMOS architecture for a micro-hotplate array. *IEEE J. Solid-State Circuits* **2007**, *42*, 441–450. [CrossRef]

127. Loutfi, A.; Coradeschi, S.; Mani, G.K.; Shankar, P.; Rayappan, J.B.B. Electronic noses for food quality: A review. *J. Food Eng.* **2015**, *144*, 103–111. [CrossRef]

![micromachines logo] *micromachines*

MDPI

Article

A 0.35-*μ*m CMOS-MEMS Oscillator for High-Resolution Distributed Mass Detection

Rafel Perelló-Roig, Jaume Verd *, Joan Barceló, Sebastià Bota and Jaume Segura

System Electronic Group (Physics Department), Universitat de les Illes Balears, Palma 07122 (Balearic Islands), Spain; rafel.perello@uib.es (R.P.-R.); j.barcelo@uib.es (J.B.); sebastia.bota@uib.es (S.B.); jaume.segura@uib.es (J.S.)
* Correspondence: jaume.verd@uib.es; Tel.: +34-971-172006

Received: 13 July 2018; Accepted: 20 September 2018; Published: 22 September 2018

Abstract: This paper presents the design, fabrication, and electrical characterization of an electrostatically actuated and capacitive sensed 2-MHz plate resonator structure that exhibits a predicted mass sensitivity of ~250 pg·cm^{-2}·Hz^{-1}. The resonator is embedded in a fully on-chip Pierce oscillator scheme, thus obtaining a quasi-digital output sensor with a short-term frequency stability of 1.2 Hz (0.63 ppm) in air conditions, corresponding to an equivalent mass noise floor as low as 300 pg·cm^{-2}. The monolithic CMOS-MEMS sensor device is fabricated using a commercial 0.35-μm 2-poly-4-metal complementary metal-oxide-semiconductor (CMOS) process, thus featuring low cost, batch production, fast turnaround time, and an easy platform for prototyping distributed mass sensors with unprecedented mass resolution for this kind of devices.

Keywords: MEMS resonators; mass sensors; CMOS-MEMS; pierce oscillator

1. Introduction and Motivation

Micro-/nanoelectromechanical systems (M/NEMS) resonators have been extensively proposed for the detection of small concentrations of analyte molecules in a gaseous solution (e.g., detection of volatile organic compounds (VOCs) through gravimetric sensing where a shift in resonance frequency is obtained in response to an added mass over the resonator). In this way, a reduction of the resonator size provides a mass sensitivity increase since the relative mass change is bigger, and in general the resonance frequency also increases. Extremely high mass sensitivity has been reported using submicrometer and nanometer scale resonators (i.e., cantilevers and CC-beams) [1–4]. Such beam-shaped resonators are the best candidates for punctual mass detection providing also an intrinsically high spatial resolution [5]. However, in applications requiring distributed mass sensing (e.g., gas detection), a relative small device area or high beam length to width ratio may represent a drawback when a subsequent resonator surface functionalization step (e.g., by ink jet, spray coat, dip cast, dip pen, or chemical deposition) is required [6,7].

In this work, we design and fabricate an alternative resonator topology based on a plate supported by four fixed-guided beams using a commercial 0.35-μm complementary metal-oxide-semiconductor (CMOS) technology. Following the fabrication technique reported in previous works, here we demonstrate the feasibility of increasing almost two orders of magnitude the resonator capture area over previous cc-beam resonators while not only preserving the final device mass resolution (per area), but enhancing it. Therefore, the predicted minimum mass change per unit area that the sensor oscillator can detect is 300 pg·cm^{-2}, which is, as far as we know, the best value reported in the literature for a monolithically integrated CMOS-MEMS device.

The paper is structured as follows: Section 2 deals with the description of the oscillator device, while in Section 3 we derive an analytical model where the resonator design parameters consider both its mass sensitivity and motional resistance with enough accuracy. In Section 4 the monolithic fabrication approach is described, and the electrical characterization results are reported. Finally, a comparative analysis with the state-of-the-art and conclusions are included in Section 5.

2. Device Description

The resonator device consists of a lateral moving plate supported by four fixed-guided beams, and two electrodes, one for electrostatic actuation, and the other for capacitive readout as illustrated in Figure 1. The top metal layer available in the commercial 0.35-μm CMOS technology is used as the physical layer to fabricate these mechanical structures [8] having an equivalent thickness (t) of 850 nm, approximately. The proposed resonator structure increases the capture area compared to the beam shaped resonators developed in previous works [9] meant for punctual mass sensing. Four anchored beams are included in the design to provide stability and to promote lateral vibration when excited by the electrodes. As further addressed in Section 3, the design parameters are chosen to optimize the distributed mass sensitivity while keeping a capture area large enough. The resonator plate area, or sensor capture area, is 41 μm (l_p) × 10.2 μm (w_p). Each beam length (l_b) is 10 μm and the width (w_b) is 0.8 μm. The gap between the resonator and the electrodes is determined by the minimum metal layer spacing allowed by the technology (s = 0.6 μm).

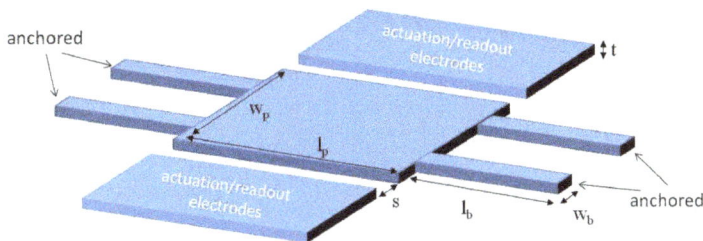

Figure 1. Schematic showing resonator dimension parameters.

The adoption of a single CMOS layer is inherently superior in mass sensitivity than the option of a stack of layers, given its lower mass. However, this implies also a lower capacitive coupling between the driver and the resonator constraining the resonator motional current detection, as detailed in next section. To overcome these issues, an on-chip high-sensitivity circuit has been integrated.

Moreover, to obtain a feasible system-on-chip, not only the readout pre-amplifier is integrated but also a full oscillator circuit scheme based on a modified Pierce topology is included, as shown in Figure 2, obtaining a quasi-digital output signal when working in closed-loop mode. The main challenge of adapting an oscillator circuit for a CMOS-MEMS resonator is its very large equivalent motional resistance (usually in the MΩ range) that must be compensated by the sustaining amplifier to enable the self-sustaining oscillator performance. The Pierce oscillator topology is used in this case, since it is superior to a transresistance amplifier in terms of the oscillator noise figure when high transimpedance gains are used [9], like that required in this work. The reason is due to the fact that most of the gain is provided by a noiseless capacitive input element rather than a lossy resistive element. In this case, the circuit exhibits a transimpedance gain of 11 MΩ and an input-referred current noise of 87 fA·Hz$^{-1/2}$ at 6 MHz.

In addition to the self-excited oscillator mode (closed-loop), the system has been designed to allow its operation also in open-loop mode (illustrated as a switch in Figure 2), thus enabling a way to characterize and test the proper resonator behavior.

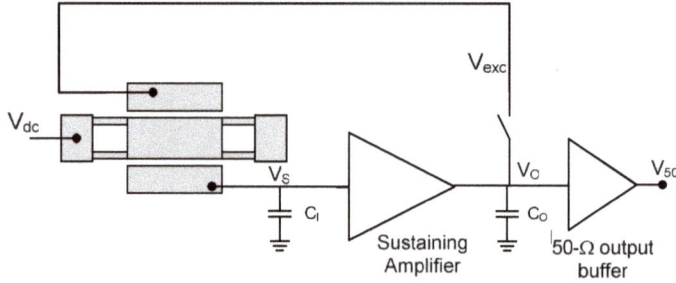

Figure 2. Conceptual circuit schematic of the lateral moving plate resonator into a Pierce oscillator topology.

3. Analytical Modelling

An approximate analytical model of the capacitive plate resonator is used in this work to estimate the resonance frequency, mass sensitivity, and electromechanical motional resistance, thus enabling a proper and simple design of the MEMS parameters. The system is conceived as a lumped mass-spring-damper model with a motion equation governing the mechanical resonator dynamics in one dimension (lateral in our case), as given by:

$$m_{\text{eff}}\ddot{x} + \gamma\dot{x} + kx = F_{\text{exc}} \tag{1}$$

where $\dot{x}$ and $\ddot{x}$ represent the first and second time derivative of the position variable x, m_{eff} is the resonator effective mass, γ is the damping coefficient, k is the linear spring constant of the system, and F_{exc} is the net electrostatic force. The system resonance frequency is obtained as $f_{\text{res}} = (k/m_{\text{eff}})^{1/2}$.

The spring constant expression of each single fixed-guided beam is [10]:

$$k_b = \frac{Ew_b^3 t}{l_b^3} \tag{2}$$

where E is the Young modulus of the structure material (top metal layer in this case). For the plate supported by four fixed-guided beams, the equivalent spring constant is:

$$k_t = 4\frac{Ew_b^3 t}{l_b^3} \tag{3}$$

The total dynamic resonator mass is calculated considering that the plate is completely rigid compared to the beams. Therefore, the system mass can be expressed in terms of the entire plate mass and the effective mass contribution of each guided-beam [11]:

$$m_{\text{eff}} = \rho l_p w_p t + 4 \times 0.37\rho l_b w_b t \tag{4}$$

Using the previous equations and assuming that the mass of the platform is larger than the one of the beams, the resonance frequency is found to be proportional to the resonator design parameters as $\sqrt{\frac{w_b^3}{l_p w_p l_b^3}}$. The resonator mass sensitivity can be expressed in terms of last parameters as:

$$S_m = -\frac{2 \times m_{\text{eff}}}{f_o} \tag{5}$$

The negative sign indicates that an added mass over the resonator results in a decrease of its resonance frequency. For distributed mass deposition (e.g., mass sensing in a gaseous solution), the sensitivity per unit of area is defined as:

$$S_{\frac{m}{a}} = -\frac{S_m}{A_{\text{eff}}} \tag{6}$$

From the last equations, it can be easily deduced that the distributed mass sensitivity is proportional to the resonator dimensions as $l_p w_p \sqrt{\frac{l_b^3}{w_b^3}} t$.

On the other hand, from the point of view of the electrostatic transduction efficiency or capacitive readout performance, the resonator is required to exhibit the minimum possible motional resistance defined as:

$$R_m = \frac{\sqrt{k_t m_{\text{eff}}}}{Q\eta^2} \tag{7}$$

where Q is the resonator quality factor related to the system damping. Considering the electrode-resonator interface as a parallel plate variable capacitor, the electromechanical transduction factor η is given by:

$$\eta = V_{\text{dc}}\frac{C_o}{s} = V_{\text{dc}}\frac{\varepsilon_o l_p t}{s^2} \tag{8}$$

and therefore, the motional resistance can be also related to the resonator dimensions in this case as $\sqrt{\frac{w_b^3 w_p}{l_b^3 l_p^3} \frac{s^4}{t}}$.

For a proper performance of the resonator in terms of mass sensitivity, but also of capacitive reading capabilities, it is required that both S_m and R_m resonator parameter values to be as small as possible. As deduced from the last equations, there is a trade-off between both parameters. While the mass sensitivity can be improved by decreasing the resonator mass and increasing both its capture area and resonance frequency, the motional resistance is reduced by increasing the resonator mass and lowering its resonance frequency. Such a trade-off is noticed mainly in terms of the guided-beams dimensions (l_b and w_b). In this sense, the resonator has been designed to exhibit the best mass sensitivity while exhibiting a motional resistance below the transimpedance gain of the CMOS sustaining circuit. Such gain must overcome the motional resistance to enable a self-sustaining oscillator operation [9]. Moreover, the frequency must be also constrained to the CMOS amplifier specifications. Figure 3 shows the dependency of the circuit transimpedance gain used in this work with the operating frequency. Since in the Pierce configuration the sensed current is integrated through the capacitance at the sustaining amplifier input (C_i in Figure 2), the amplifier transimpedance gain decreases with frequency.

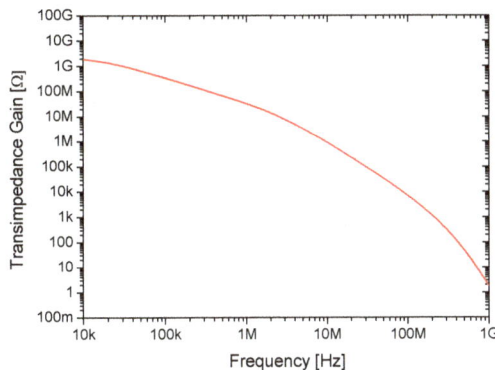

Figure 3. Transimpedance gain versus frequency of the sustaining amplifier circuit.

Therefore, the beam width (w_b) is chosen as the minimum available by technology since a wider beam highly increases both the frequency and the motional resistance. The platform width and length (w_p and l_p) are chosen to fulfill the capture area requirements, with the length larger than the width, to avoid increasing the motional resistance unnecessarily. Thus, the parameter available for the design flow is the beam length (l_b). Finally, we also set the gap between the resonator and the electrode (s) as the minimum available by the technology to minimize the motional resistance as well. In any case, large motional resistance values are still obtained since the use of a single top metal layer with a relatively small thickness and a relatively large minimum distance as indicated in Section 2.

This analytical model has been used to design the resonator parameters according to the values indicated in Section 2. Additionally, we ran extensive finite element modeling (FEM) simulations using COMSOL Multiphysics to validate the model parameters prior to fabrication. Comparison of the accurate finite element modeling (FEM) values to the analytically derived parameters revealed that a high accuracy of such a simple model is good enough to design the resonator parameters and predict its main performance as a sensor device. Considering that the spring constant may usually vary by 10%–20% due to process variations in conventional MEMS fabrication techniques [10], the results detailed in Table 1 are absolutely acceptable.

Table 1. Main resonator parameters comparison obtained from finite element modeling (FEM) simulations and using the approximated analytical solution. A mass density ρ = 3000 kg/m^3 and a Young's modulus E = 131 GPa have been assumed for the complementary metal-oxide-semiconductor (CMOS) top metal layer to compute the parameters.

Parameter	FEM	Analytical	Error (%)
Resonance frequency, f_o (MHz)	2.124	2.296	−7.5
Linear stiffness, k (N/m)	199.8	228	−12.3
Mass sensitivity, $S_{\frac{m}{a}}$ (pg·cm^{-2}·Hz^{-1})	246.8	228.5	8.0

4. Fabrication and Electrical Characterization

The MEMS device was completely defined along the commercial 0.35-μm CMOS process by using the top metal layer available in the technology used [8]. The silicon oxide underneath the resonator was used as the sacrificial layer that was removed after the standard CMOS process by means of a one-step maskless wet etching. This construction scheme provides an easy monolithic integration of the mechanical resonator with the CMOS circuitry (see Figure 4a). In this case, various openings were defined in the plate structure to ease the wet etching process (Figure 4b).

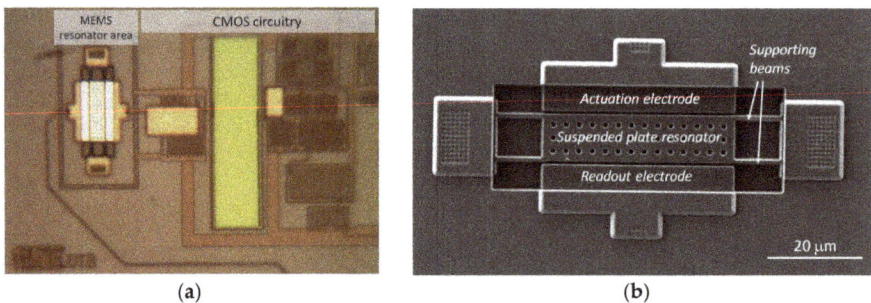

(a)   (b)

Figure 4. Fabricated device in a complementary metal-oxide-semiconductor (CMOS) 0.35-μm commercial technology constituted by a plate resonator integrated monolithically with CMOS circuitry: (**a**) Optical image of the overall CMOS-MEMS oscillator circuit; (**b**) SEM image of the metal suspended plate resonator.

The CMOS-MEMS device has been characterized both in open-loop and closed-loop mode in air conditions (atmospheric pressure and ambient temperature). Open-loop measurements have been performed at a low excitation power (−30 dBm) to operate in the resonator linear region, enabling the extraction of the linear electromechanical parameters. Figure 5 shows the measured electromechanical system transmission response magnitude obtained for various resonator bias voltages V_{dc}. The resonance frequency of the MEMS resonator is around 1.9 MHz and its quality factor is $Q = 176$. On the other hand, it is noticed that the spring softening effect (the decrease of resonant frequency when the bias voltage increase) is not marked in contrast to the one exhibited by clamped-free resonators. From such behavior (Figure 5b), the pull-in voltage of the device is predicted to be as high as 380 V.

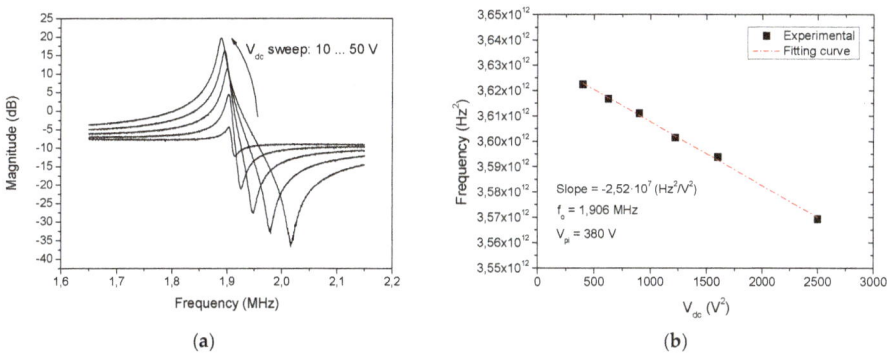

(a)   (b)

Figure 5. Electrical characterization of the suspended plate resonator with on-chip 0.35-μm CMOS readout circuit in open-loop configuration: (**a**) Measured frequency response (magnitude) for different resonator bias voltages (V_{dc}) in air conditions; (**b**) Plot of the resonance frequency dependency versus the applied bias voltage.

The oscillator output signal behavior, when working in self-excited mode (closed-loop measurements), is depicted in Figure 6. The oscillator works properly with biasing voltages over 25 V. For a $V_{dc} = 27$ V, the measured oscillator generates a 1.947 MHz signal with a peak-to-peak amplitude beyond 500 mV. In resonant sensing, where the measurement is performed by tracking the MEMS resonance frequency variation, the Allan deviation parameter of the oscillator frequency becomes a key parameter widely used to assess the short-term stability of oscillators and to predict, in this case, the short-term resolution of the device (e.g., mass resolution). In this work, a precise frequency counter (CNT-90, Pendulum, Banino, Poland) has been used to measure the Allan deviation of the CMOS-MEMS oscillator in a range from 1 ms to 1 s integration times (Figure 7). The integration time is equivalent, in the time domain, to the measurement bandwidth in the frequency domain used in other instruments like spectrum analyzers.

The corresponding surface mass limit of detection (SMLOD) is also provided in Figure 7. This parameter accounts for the minimum mass change per unit area that the CMOS-MEMS device can detect. It has been computed from the predicted mass sensitivity times of the measured Allan deviation value. The minimum Allan deviation value is obtained for an integration time of 100 ms being as low as 1.2 Hz or 0.63 ppm.

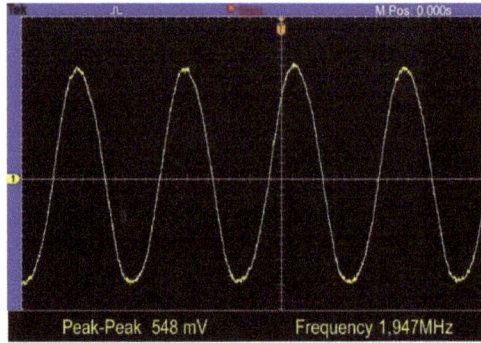

Figure 6. Time-domain oscillator output signal measured for V_{dc} = 27 V.

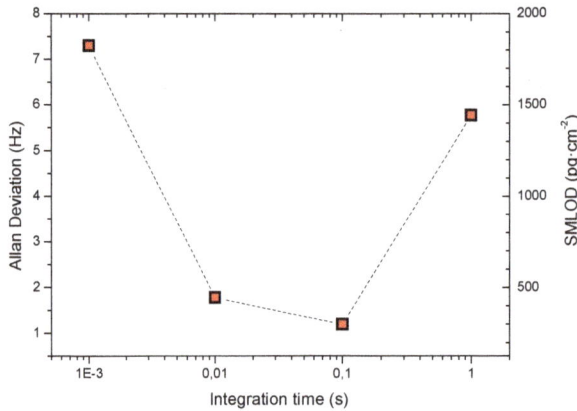

Figure 7. Allan deviation as a function of integration time measured in air conditions (atmospheric pressure and ambient temperature). The corresponding surface mass limit of detection (SMLOD) for each integration time is indicated in the right y-axis.

5. Comparative and Conclusions

The plate resonator reported in this work has a capture area ~40 times higher than featured by the cc-beam resonator used in previous works [12], and is thus more suitable for distributed mass sensing applications as introduced in Section 1. On the other hand, its higher mass and lower resonant frequency results in a predicted mass sensitivity per unit area to be ~7 times worse (250 $pg \cdot cm^{-2} \cdot Hz^{-1}$). However, the better frequency stability of the device results in a predicted surface mass limit of detection (SMLOD) to be as low as 300 $pg \cdot cm^{-2}$, and is thus better than obtained for the cc-beam resonator in air conditions. In addition, the plate resonator operates at a lower bias voltage. In our opinion, the better oscillator frequency stability is related to a higher mechanical linearity and quality factor that exhibits this type of MEMS structure compared to the single cc-beam previously used in [12]. Being an open topic, we are currently investigating in this direction to corroborate such assumption.

The predicted SMLOD value and the expected mass sensitivity per unit area obtained in this work is also the best compared to the state-of-the-art for monolithic CMOS-MEMS oscillators (with on-chip feedback circuitry) operating in air conditions, as shown in Table 2. Only a few works addressing CMOS compatible MEMS mass sensors report smaller SMLOD values [13,14]. However, the better frequency stability in [14] is obtained thanks to the use of external off-chip feedback circuitry based on a phase locked loop (PLL). In [13], they also used an off-chip PLL, and the better mass sensitivity is inherent to the use of a nano-scale cantilever structure having a 10^2 times smaller capture area. On the other hand, comparison with non-CMOS/MEMS resonators reveals that only some works based on piezoelectric transduction and fabricated using specific nanofabrication techniques exhibit better performance in terms of SMLOD (e.g., in References [15,16]).

Table 2. State-of-the-art of CMOS-MEMS resonators as distributed mass sensors operating in air conditions.

Ref.	Sensitivity ($pg \cdot cm^{-2} \cdot Hz^{-1}$)	Capture Area (cm^2)	Frequency Stability (Hz)	SMLOD ($pg \cdot cm^{-2}$)	Integrability
This work	250 *	4.2×10^{-6}	1.2	300 *	On-chip oscillator
[12]	34 *	1.1×10^{-7}	15	510 *	On-chip oscillator
[13]	0.10 *	1.2×10^{-9}	20	2 *	Off-chip PLL
[14]	2300	2.2×10^{-4}	0.08	190	Off-chip PLL
[17]	61,000	2.3×10^{-4}	0.03	1800	On-chip oscillator
[18]	240,000	1×10^{-4}	~1	240,000	On-chip oscillator

* Predicted value.

As a main conclusion, in this work we have experimentally demonstrated the feasibility of constructing a fully integrated and fully standard CMOS fabrication process compatible MEMS-based oscillator with a surface beyond 60 times of single cc-beams with a superior SMLOD in air conditions. Given its large area and high mass sensitivity per unit area, this device is a perfect candidate for applications requiring selective compound detection in gas environments where functionalization steps are mandatory. The higher capture area ensure proper adherence of the subsequent chemical species, thus avoiding the contour-dominated effect intrinsic disadvantages of narrow cc-beams.

Finally, the full CMOS/MEMS integration of the mechanical and electronic circuitry and the quasi-digital output provided by the oscillator makes this solution a perfect candidate for low cost system-on-chip or Lab-on-chip applications.

Author Contributions: Conceptualization—R.P.-R., J.V. and J.S.; Formal analysis—J.B.; Investigation—R.P.-R. and J.V.; Methodology—R.P.-R. and J.S.; Resources—S.B.; Supervision—J.V. and J.S.; Writing review & editing—R.P.-R., J.V. and J.S.

Funding: This work has been supported by the Spanish Ministry of Economy and Competitiveness under project TEC2017-88635-R (AEI/FEDER, UE).

Acknowledgments: R. Perelló-Roig thanks his grant FPU-16/01758 from the Spanish Ministry of Education, Culture and Sport.

Conflicts of Interest: The authors declare no conflict of interest.

References

1. Ouerghi, I.; Philippe, J.; Duraffourg, L.; Laurent, L.; Testini, A.; Benedetto, K.; Charvet, A.M.; Delaye, V.; Masarotto, L.; Scheiblin, P.; et al. High performance polysilicon nanowire NEMS for CMOS embedded nanosensors. In Proceedings of the 2014 IEEE International Electron Devices Meeting, San Francisco, CA, USA, 15–17 December 2014.
2. Li, M.; Tang, H.X.; Roukes, M.L. Ultra-sensitive NEMS-based cantilevers for sensing, scanned probe and very high-frequency applications. *Nat. Nanotechnol.* **2017**, *2*, 114–120. [CrossRef] [PubMed]
3. Ilic, B.; Craighead, H.G.; Krylov, S.; Senaratne, W.; Ober, C.; Neuzil, P. Attogram detection using nanoelectromechanical devices. *Appl. Phys. Lett.* **2004**, *85*, 2604–2606. [CrossRef]
4. Lavrik, N.V.; Sepaniak, M.J.; Datskos, P.G. Cantilever transducers as a platform for chemical and biological sensors. *Rev. Sci. Instrum.* **2004**, *85*, 2229–2253. [CrossRef]
5. Arcamone, J.; Sansa, M.; Verd, J.; Uranga, A.; Abadal, G.; Barniol, N.; van den Boogaart, M.; Brugger, J.; Pérez-Murano, F. Nanomechanical mass sensor for spatially resolved ultrasensitive monitoring of deposition rates in stencil lithography. *Small* **2009**, *5*, 176–180. [CrossRef] [PubMed]
6. Bedair, S.S.; Fedder, G.K. Polymer wicking to mass load cantilevers for chemical gravimetric sensors. In Proceedings of the 13th International Conference on Solid-State Sensors, Actuators and Microsystems, Seoul, Korea, 5–9 June 2005.

7. Voiculescu, I.; Zaghloul, M.E.; McGill, R.A.; Houser, E.J.; Fedder, G.K. Electrostatically actuated resonant microcantilever beam in CMOS technology for the detection of chemical weapons. *IEEE Sens. J.* **2005**, *5*, 641–647. [CrossRef]
8. Verd, J.; Uranga, A.; Teva, J.; Lopez, J.L.; Torres, F.; Esteve, J.; Abadal, G.; Pérez-Murano, F.; Barniol, N. Integrated CMOS-MEMS with on-chip readout electronics for high-frequency applications. *IEEE Electron Device Lett.* **2006**, *27*, 495–497. [CrossRef]
9. Verd, J.; Uranga, A.; Abadal, G.; Teva, J.L.; Torres, F.; Lopez, J.; PÉrez-Murano, F.; Esteve, J.; Barniol, N. Monolithic CMOS MEMS oscillator circuit for sensing in the attogram range. *IEEE Electron Device Lett.* **2008**, *29*, 146–148. [CrossRef]
10. Kaajakari, V. *Practical MEMS: Design of Microsystems, Accelerometers, Gyroscopes, RF MEMS, Optical MEMS, and Microfluidic Systems*; Small Gear Publishing: Las Vegas, NV, USA, 2009.
11. Wai-Chi, W.; Azid, A.A.; Majlis, B.Y. Formulation of stiffness constant and effective mass for a folded beam. *Arch. Mech.* **2010**, *62*, 405–418.
12. Verd, J.; Sansa, M.; Uranga, A.; Perez-Murano, F.; Segura, J.; Barniol, N. Metal microelectromechanical oscillator exhibiting ultra-high water vapor resolution. *Lab Chip* **2011**, *11*, 2670–2672. [CrossRef] [PubMed]
13. Arcamone, J.; Dupré, C.; Arndt, G.; Colinet, E.; Hentz, S.; Ollier, E.; Duraffourg, L. VHF NEMS-CMOS piezoresistive resonators for advanced sensing applications. *Nanotechnology* **2014**, *25*, 1–9. [CrossRef] [PubMed]
14. Liu, T.Y.; Chu, C.C.; Li, M.H.; Liu, C.Y.; Lo, C.Y.; Li, S.S. CMOS-MEMS thermal-piezoresistive oscillators with high transduction efficiency for mass sensing applications. In Proceedings of the 19th International Conference on Solid-State Sensors, Actuators and Microsystems (Transducers), Kaohsiung, Taiwan, 18–22 June 2017.
15. Ivaldi, P.; Abergel, J.; Matheny, M.H.; Villanueva, L.G.; Karabalin, R.B.; Roukes, M.L.; Andreucci, P.; Hentz, S.; Defaÿ, E. 50 nm thick AlN film-based piezoelectric cantilevers for gravimetric detection. *J. Micromech. Microeng.* **2011**, *21*, 085023. [CrossRef]
16. Li, M.; Myers, E.B.; Tang, H.X.; Aldridge, S.J.; McCaig, H.C.; Whiting, J.J.; Simonson, R.J.; Lewis, N.S.; Roukes, M.L. Nanoelectromechanical resonator arrays for ultrafast, gas-phase chromatographic chemical analysis. *Nano Lett.* **2010**, *10*, 3899–3903. [CrossRef] [PubMed]
17. Lange, D.; Hagleitner, C.; Hierlemann, A.; Brand, O.; Baltes, H. Complementary metal oxide Semiconductor cantilever arrays on a single chip: Mass-sensitive detection of volatile compounds. *Anal. Chem.* **2002**, *74*, 3084–3095. [CrossRef] [PubMed]
18. Bedair, S.S.; Fedder, G.K. CMOS MEMS oscillator for gas chemical detection. In Proceedings of the SENSORS, 2004 IEEE, Vienna, Austria, 24–27 October 2004.

micromachines

MDPI

Article

Array of Resonant Electromechanical Nanosystems: A Technological Breakthrough for Uncooled Infrared Imaging

Laurent Duraffourg [1,2,*], **Ludovic Laurent** [1,2], **Jean-Sébastien Moulet** [1,2], **Julien Arcamone** [1,2] and **Jean-Jacques Yon** [1,2]

[1]   Université Grenoble Alpes, F-38000 Grenoble, France; ludovic.laurent@mirsense.com (L.L.);
       jean-sebastien.moulet@cea.fr (J.-S.M.); julien.arcamone@cea.fr (J.A.); jean-jacques.yon@cea.fr (J.-J.Y.)
[2]   CEA, LETI, Minatec Campus, F-38054 Grenoble, France
*   Correspondence: laurent.duraffourg@cea.fr; Tel.: +33-43-878-2915

Received: 7 July 2018; Accepted: 6 August 2018; Published: 14 August 2018

Abstract: Microbolometers arethe most common uncooled infrared techniques that allow 50 mK-temperature resolution to be achieved on-scene. However, this approach struggles with both self-heating, which is inherent to the resistive readout principle, and 1/f noise. We present an alternative approach that consists of using micro/nanoresonators vibrating according to a torsional mode, and whose resonant frequency changes with the incident IR-radiation. Dense arrays of such electromechanical structures were fabricated with a 12 µm pitch at low temperature, allowing their integration on complementary metal-oxide-semiconductor (CMOS) circuits according to a post-processing method. H-shape pixels with 9 µm-long nanorods and a cross-section of 250 nm × 30 nm were fabricated to provide large thermal responses, whose experimental measurements reached up to 1024 Hz/nW. These electromechanical resonators featured a noise equivalent power of 140 pW for a response time of less than 1 ms. To our knowledge, these performances are unrivaled with such small dimensions. We also showed that a temperature sensitivity of 20 mK within a 100 ms integration time is conceivable at a 12 µm pitch by co-integrating the resonators with their readout electronics, and suggesting a new readout scheme. This sensitivity could be reached short-term by depositing on top of the nanorods a vanadium oxide layer that had a phase-transition that could possibly enhance the thermal response by one order of magnitude.

Keywords: nano resonator; nano-system array; uncooled IR-bolometer

1. Introduction

Microelectromechanical systems (MEMS) are either used according to a static mode (accelerometers, micro-mirrors, radio frequency (RF) switches, bimorph structures) or rather in a dynamic way, when the harmonic response of the system is requested. Although quartz crystals were first studied (and continue to be widely used as time references), silicon is most likely the most widely used material for MEMS and NEMS (nano-electromechanical systems) due to its excellent mechanical properties. Indeed, advances in microelectronics on silicon wafers (increasingly large diameters—the reliability and reproducibility of manufacturing methods, silicon on insulator (SOI) substrates) have strongly enabled the rise of silicon-based MEMS/NEMS. Beyond silicon, alternative materials are nowadays already industrialized, such as lead zirconate titanate (PZT) and AlN, or envisaged such as GaAs, graphene, or aluminum, generally deposited on silicon substrates (in particular for economic reasons). Finally, the intrinsic compatibility of the manufacturing of silicon or silicon-based materials with a complementary metal-oxide-semiconductor (CMOS) integration ("above-integrated circuit (IC)" approach or co-integration approach) is an undeniable asset in order to realize a high-signal-to-noise ratio (SNR) system including an actuator/detector and a compact readout circuit that is affordable and energy efficient.

The small size of NEMS makes them particularly sensitive to their external environments, while keeping very good frequency stability. In particular, silicon-based nanoresonators have already demonstrated their formidable potential in their number of application domains. Nanoresonators are excellent physical sensors, especially for measuring forces [1,2] or mass [3,4], for gas detection [5,6], but also for measuring the temperature [7–10]. The ultimate sensitivity of the order of the yoctogram (10^{-24} g—mass of a proton) has even been demonstrated with single-wall carbon nanotubes (CNT) [11,12]. More generally, nanoresonators can extend proteomics to high mass biomolecules (such as complexes of proteins or viruses) [3,13]. There is also intense activity in the NEMS community related to the study of oscillators in their fundamental quantum mode using the reciprocal interaction of an optical micro cavity with a mechanical resonator [14–16]. Examples of mechanical resonator applications are plethoric: particle counting in a fluid medium [17,18], magnetometry [19–23], actuators [24], RF filters [25], and in biology [26,27].

The measurement principle is quite simple: it consists of monitoring the frequency shift of a NEMS kept at a given vibration (at a fixed controlled amplitude) using a closed loop circuit via a "phase-locked loop" (PLL) or a self-oscillating circuit. Any change in an external physical parameter (temperature, pressure, force, acceleration etc.) or on the surface of the material (adsorption of a gas, molecule, etc.) will modify its stiffness or its mass, thus inducing a change in the resonant frequency, which is continuously measured. Figure 1, adapted from [28], illustrates the measurement principle for when a frequency shift is caused by a mass landing on top of a nanocantilever. In this case, a piezoresistive transduction was used to measure the mechanical oscillations of the cantilever (the actuation being purely capacitive [29]). Indeed, it was shown that piezoresistive detection is highly suitable at high frequency compared to the capacitive readout in a self-oscillation loop [30].

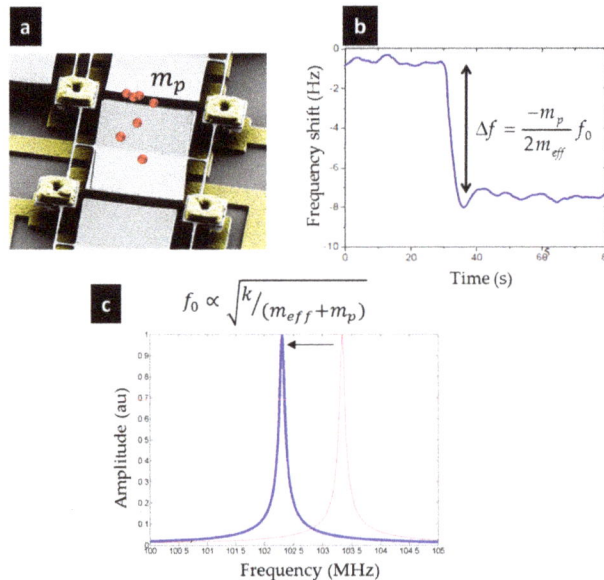

Figure 1. The principle of mass measurement (adapted from [28]). (**a**) Example of a nanoresonator on which particles have landed; (**b**) Shift in the frequency caused by the arrival of particles. Monitoring in real-time the resonance frequency allows us to deduce the amount of accreted mass; (**c**) From the spectral perspective: a shift in the spectrum toward low frequencies.

Neutral mass spectroscopy with such a system is now ongoing, and some papers have already demonstrated their interest for biomolecule analysis [3]. Beyond the ultra-sensitivity of NEMS, the overall analysis time has to be fast enough to make this technique a realistic solution. To tackle this key issue, the basic method consists of using a NEMS array for increasing the capture cross-section, and hence speeding up the analysis. In a recently published work, a NEMS array dedicated to mass sensing was realized with a frequency address, with each NEMS having a slightly different resonance frequency that labelled its position inside the array [31].

Nanoresonators can also be used for thermal sensing. Resonant NEMS arrays with a suitable driving electronics could be a way to continue to decrease the pixel pitch of the thermal imager, keeping the performance constant. Among the various current uncooled infrared focal plane array (IRFPA) technologies, the microbolometer is the most common uncooled infrared device. It operates by converting the heating of a thin suspended membrane due to IR absorption into a variation of the electrical resistance of a layer deposited on it. This layer is commonly made of a thin film of semiconductor, using mostly vanadium oxide (VOX) or amorphous silicon (a-Si), because of their high thermal sensitivity ($1/R \times dR/dT$), which is about 2–3%/K. To reach a high thermal sensitivity, the membrane has to be insulated from the substrate, and it is suspended above the readout integrated circuit (ROIC) by thin and long insulation legs. Thermal insulation as high as 2 mK/W has, for instance, been reported for microbolometers with 12 µm pixel pitch [32]. Such a development has achieved noise equivalent temperature differences (NETD) at a very low level, around 50 mK (F/1 lens, 30 Hz frame rate, 300 K background) [33,34]. At ultra-small pixel pitches (less than 12 µm), the performances could be kept to the cost of a very high thermal insulator, leading to a dramatic temperature increase of the suspended plate, which can be deleterious for the electrical properties of the sensing material [35]. Both VOX and a-Si microbolometers can be affected [36]. The direct consequence is a persistent afterimage when the scene is removing [37]. This is particularly true for when the microbolometer is exposed to a high temperature source, such as fire, explosion, or the sun, and is this sometime referred to as the "Sun Burn" effect [38]. Although a shutter-based non-uniformity correction can mitigate this effect, latent images reappear rapidly after shutter operation and persist with a very long decay time. Thus, there is a clear need for a new transduction method that can withstand high temperature exposure. In parallel, this technique has to be compatible with very large scale manufacturing for future consumer markets that require small pixel-pitch and high resolution.

In this context, we suggest a new transduction method based on high-frequency mechanical nanoresonators designed to be ultra-sensitive to IR radiation. For mid-term vision, this approach intends to replace the current thermistor-based bolometers. The frequency stability of these nanoresonators should allow for the fundamental phonon noise to be reached, and to better cope with the thermal issues, while keeping a high resolution on the scene for the small pixel sizes. In principle, the excellent frequency sensitivity of nano/micro resonators makes them perfect ultra-small thermal sensors. However, two main questions must be raised: how can an efficient electrical transduction with a tiny mechanical displacement be realized without any self-heating? Is the frame rate fast enough with such an approach to obtain a net image? This paper will provide some key insights, using a CMOS-based approach.

In a first level of answer, the transduction technique has to be properly chosen. Unlike the conclusions made in previous papers [30], capacitive transduction is preferred for removing the self-heating that results in a background signal and additional noise. Ultra-small capacitance variation is however, more delicate to read out [39], and a specific buffer circuit has to be developed, as we will show later on in the paper. Many types of transductions have been suggested over the last 10 years. A second level of the answer lies in the use of the NEMS array collectively or individually actuated, and detected by a CMOS circuit that could be placed in its close vicinity. The CMOS circuit will actuate the mechanical systems at their resonance frequencies, and perform the addressing of a single NEMS or a sub-array of NEMS to be addressed. The spatial proximity of the CMOS with the thermal sensors drastically limits the effects of electrical parasitic coupling and attenuation. Limiting these effects consequently maximizes at circuit input, both the absolute value of the useful signal, and its signal-to-background ratio (SBR), resulting in maximization of the SNR at the circuit output. To conclude, this juxtaposition has significant advantages, namely the compactness of the system, and the unparalleled electrical transduction efficiency.

2. Design and Fabrication of Electromechanical Resonator Arrays

The basic geometry of the resonator was a suspended plate that experienced a torsional vibration around a rotation axis. Several metallic electrodes were structured underneath for electrostatic actuation and capacitive measurement of the paddle displacement. These electrodes and the paddle formed a λ/4-resonant cavity centered at 8 μm for enhancing the absorption of the incident IR-radiation. The Fabrication process and the materials had to be temperature-compatible with monolithic integration, in order to simplify the manufacturing of the imagers and their integration on a CMOS readout circuit. Thus, the materials must meet three main criteria: (i) good mechanical features; (ii) low thermal conductivity; (iii) low-temperature deposition. The critical dimensions (as the width legs ensuring the rotation and the thermal insulation of the plate) had to be well controlled during the fabrication process to ensure that all pixels were functional inside the imager.

The low-temperature fabrication process was derived from classical bolometers (i.e., a deposition process <300 °C and above-IC compatible). First, a 300 nm thick AlCu layer was deposited on a silicon substrate and structured to form the transduction electrodes. A 2 μm-deposition of a polyimide layer was then formed and constituted the sacrificial layer. The latter was opened to build up the metal studs that would insure the mechanical support and the electrical connection with the electrical connections below. Two silicon nitride (SiN) layers of 10 nm encapsulated a titanium nitride (TiN) layer that acted as an electrode as well as an absorber. The TiN thickness was defined to be impedance-matched with the vacuum (Z_0 ~376 Ω), in order to obtain a direct absorption rate that was close to 50%. The λ/4-optical cavity (2 μm thick) between the aluminum–copper electrodes and the TiN layer allowed an absorption efficiency of 80% to be reached over the 8–14 μm wavelength range. Figure 2g shows the spectral absorption of such a cavity with this specific SiN/TiN/SiN/a-Si stack in this wavelength range.

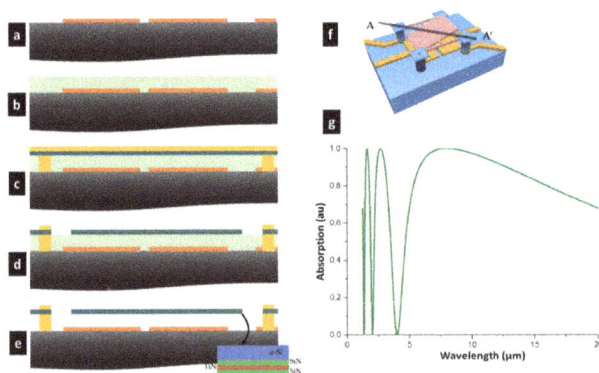

Figure 2. Synopsis of the fabrication process along the cross section AA': (**a**) Deposition of a 300 nm thick AlCu layer on a silicon substrate/strip lines/reflector wet etching; (**b**) 2 μm deposition of a polyimide layer; (**c**) Etching of the polyimide layer to build up the metal studs, and deposition of the plate material; (**d**) Definition of the plate; (**e**) Dry etching O_2 plasma release; (**f**) 3D artist view of a pixel; (**g**) Absorption spectrum of the 2 μm thick micro cavity.

The encapsulation of the TiN layer was performed for stress compensation reasons and to protect it during the release step. An amorphous silicon of 150 nm thickness (a-Si) was deposited on the top SiN to stiffen the plate. The plate had to be polarized through the legs via the thin TiN layer.

Electrostatic actuation was possible in these conditions, since no high current was required. Electrical contacts were made by opening the top layers (SiN and a-Si).

The torsional mechanical eigenmode has always been addressed in all designs of pixels so far. Indeed, the advantages of the torsional mode are threefold: (i) this mode is less sensitive to the residual axial stress that could be different from one side to the other side of an array [9,40]; (ii) the dynamic range set by the onset of nonlinearity is higher for the torsional mode compared to the flexural modes [9,40–42], since only the external fiber of the rods experiences a strain [43]; (iii) the paddle surface remains large compared to the overall resonant body and makes a capacitive actuation easy. Resonator arrays of 666 × 520 pixels, with a 12 μm-pitch were fabricated. Figure 3a shows a scanning electron microscopy (SEM) picture of a typical array. The first electromechanical tests were achieved using polarization lines structured below the pixel. This interconnection can be observed on Figure 3a, and enabled the actuation and read out of an array of 96 × 96 electromechanical pixels. An SEM zoom-in of a typical H-shape pixel is presented in Figure 3b,c. The nano-rod length was 1.5 μm for a cross-section of 250 nm × 180 nm (width × thickness). The insulation arm length was 8.6 μm. This design was the nominal version of our electromechanical pixel. Other versions were, however, realized and some of them are presented in Figure 4a–d. These alternative versions will be reviewed in the next section: they were conceived in an attempt to meet the best trade-off between an efficient thermal insulation and a large mechanical dynamic range, these two key features being antagonists. Indeed, it is interesting to look at the Equation (1) below. It shows that the thermal response is inversely proportional to the thermal conductance, *G*. The thermal insulation should be improved by increasing the lengths of the legs and the rods. At the same time, the torsional stiffness should be high enough to prevent the occurrence of quick and strong nonlinear effects. This requires quite short legs. The lengths and widths of the legs versus the rods had to be carefully chosen to find the best operating point (large linear displacement, large thermal insulation, and low driving voltages). Notice that even if the asymmetric designs were drawn to enhance the electrostatic actuation, a square shape could be realized. Along the two directions, the pitch was nevertheless kept constant at 12 μm.

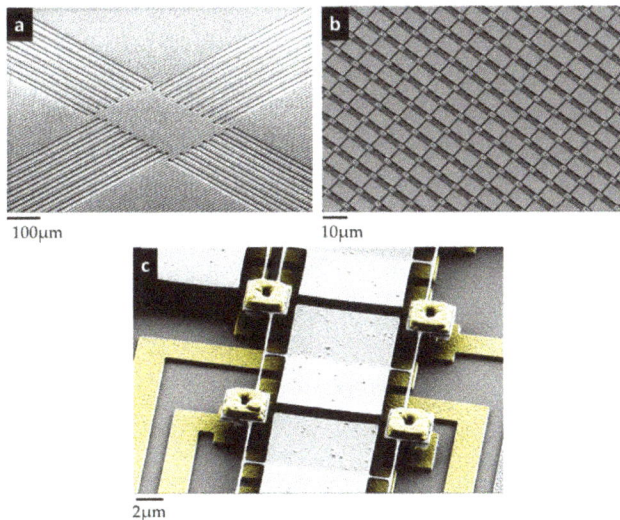

Figure 3. Scanning electron microscopy (SEM) pictures of an array of electromechanical pixels fabricated with a low-temperature process: (**a**) Large field view of an array; only the central 96 × 96 array is connected to electrical pads; pixels above the connection wires have been removed to avoid any cross-talk; (**b**) Zoom-in on the center of the array; (**c**) SEM picture of a typical H-shape pixel; Nanorod length = 1.5 μm, width = 250 nm and thickness = 180 nm (insulation arm length = 8.6 μm).

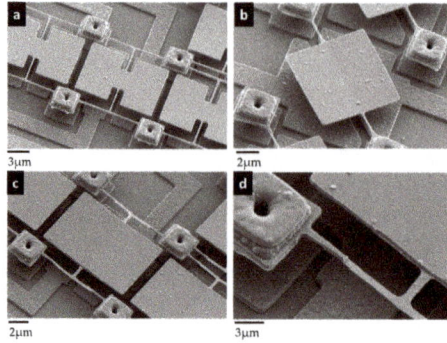

Figure 4. SEM pictures of alternative versions derived from the nominal design: (**a**) Butterfly-shaped pixel with longer rods; (**b**) Simple pixel without insulating legs; (**c**) H-shape pixel with thinner nanorods for enhancing the thermal insulation; rod-thickness = 30 nm; (**d**) Zoom of the legs (Figure 4c) attached to a stud that acts as a mechanical anchor and that provides electrical contact with the lines underneath.

For the sake of clarity, a short introduction to the key mechanisms and noise sources is presented below. We did not aim at detailing the thermo-electromechanical equations that described the overall interactions between the mechanics and the IR-light. Rather, we gave key expressions for catching up this approach that may constitute a new paradigm in the field of IR-imaging. The overall measurement system, including a single pixel, is depicted in Figure 5. The expressions of parameters shown in this figure are detailed step-by-step below for a comprehensive vision. The further expressions are appropriate under a small-displacement (small deflection angle) assumption. The polarization of the pixel is set to keep the angular vibration in its linear range at the chosen torsional resonance frequency. V_B is the bias voltage applied on the paddle (through the studs), and V_{AC} is the sinusoidal polarization applied on the actuation electrode through the capacitance Ca. This signal can be applied with an external RF-source, in particular for the first electromechanical characterizations, but can come from the feedback loop in the case of a closed loop. The actuation frequency f is swept to measure the electromechanical response and the resonance frequency f_0.

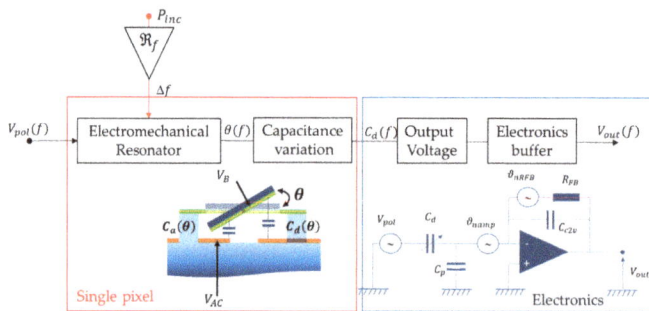

Figure 5. Synopsis of the open-loop measurement chain: The red box corresponds to a single electromechanical pixel that translates the incident IR-radiation P_{inc} on the scene into a resonance frequency shift; the blue box corresponds to the close-by electronics that convert the mechanical oscillations into an electrical signal; V_{pol} is the polarization of the pixel; $\theta(f)$ is the angular oscillation of the paddle around the rods; $Cd(f)$ is the induced capacitance variation used to read out the signal; $V_{out}(f)$ is the output signal supplied by the buffer. C_p is the total capacitance due to amplifier input capacitance, and parasitic capacitances between the electrical connections and the ground.

Basically, the electromechanical pixel converts the incident IR-optical power P_{inc} into a resonance frequency shift Δf according to a sensitivity Rf, which depends on both the thermal conductance of the paddle insulation (through insulation legs between the torsional rods and the plate) and the temperature coefficient of frequency:

$$\Delta f = \frac{\alpha_T f_0 \beta \eta}{G|1 + j2\pi v \tau_{th}|} P_{inc} = f_0 R_f P_{inc} \tag{1}$$

where $\tau_{th} = C/G$ is the thermal time constant of the sensor, $C = \left(\frac{\partial U}{\partial T}\right)_V$, the thermal capacitance at constant volume, G is the thermal conductance, α_T is the temperature coefficient of frequency (TCF) (typically -60 ppm/°C for silicon), β is the pixel fill factor, η is the bolometer absorption, f_0 is the resonance frequency, and v the frame rate of the electronic readout. The thermal conductance is mainly due to the thermal conductance of heat through the legs. The other sources of thermal leaks—radiative and heat conductance through air—are negligible.

The capacitance variations can be calculated from geometrical considerations. After cumbersome mathematical manipulation, the final expressions can be approximated as:

$$C_a(\theta) \approx -C_0 \frac{\theta_{max}}{\theta} \ln\left(1 - \frac{\theta}{\theta_{max}}\right) \tag{2}$$

$$C_d(\theta) \approx C_0 \frac{\theta_{max}}{\theta} \ln\left(1 + \frac{\theta}{\theta_{max}}\right) \tag{3}$$

$C_0 = \varepsilon_0 \frac{L_p W_p / 2}{g}$ and $\sin(\theta_{max}) = \frac{g}{W_p/2}$. C_0 and θ_{max} are respectively the capacitance value at rest, and the maximum deflection angle. The deflection angle is directly computed from the dynamic equation:

$$J\ddot{\theta} + b\dot{\theta} + \kappa\theta = T_e \Delta f = \frac{\alpha_T f_0 \beta \eta}{G|1 + j2\pi v \tau_{th}|} P_{inc} = f_0 R_f P_{inc}$$

$$T_e = \frac{1}{2}\frac{dC_a}{d\theta}V_{pol}^2 + \frac{1}{2}\frac{dC_d}{d\theta}V_B^2 \tag{4}$$

$$J = \frac{M_p W_p^2}{12}; \ \kappa = \frac{2G I_r}{L_r} \text{ and } G = \frac{E}{2(1+v)}$$

T_e is the electrostatic torque. J is the moment of inertia of the paddle, assuming that the inertia moment of rods is negligible. k, G, and I_r are respectively the rod torsional stiffness, the shear modulus, and the torsional quadratic moment of the rectangular suspended rods ($I_r = w_r t_r^3 \left(\frac{1}{3} - 0.21\frac{t_r}{w_r}\left(1 - \frac{1}{12}\frac{t_r^4}{w_r^4}\right)\right)$, $w_r > t_r$.). E and v are the equivalent Young modulus and Poisson ratio of the stack.

Under the assumption of a linear regime (and small deflection amplitude of the paddle), the angle can be rewritten in the Fourier space:

$$\theta(f) = \frac{T_e}{J}\frac{1}{f_0^2 - f^2 + j\frac{ff_0}{Q}} \tag{5}$$

Table 1 presents the values of the main features of Equations (1)–(5) for our typical electromechanical pixel. Some parameters are compared with data from literature.

Table 1. Key parameters presented in the Equations (1)–(5) for our device compared with an advanced resistive bolometer and microelectromechanical systems (MEMS) bolometer: temperature sensitivity corresponds to $\frac{1}{f} \times \partial f \partial T$ for a resonant thermal sensor and $\frac{1}{R} \times \partial R \partial T$ for a resistive one.

Electromechanical & Thermal Features	This Work (Figure 3c)	Bolometer [32]	Resonant MEMS [9]
Maximal Angle θ_{max} (°)	21	N.A	-
Inertial Moment J (kg.m^2)	3.9×10^{-25}	N.A	1.5×10^{-27}
Torsional stiffness κ (N.m)	1.8×10^{-11}	N.A	6.8×10^{-13}
Resonant Frequency (MHz)	1.1	N.A	-
Onset of Nonlinearity θ_c(°) (This value is computed by solving a nonlinear dynamic equation [44].)	13.5	N.A	-
Quality Factor Q	1800	N.A	1555
Capacitance at Rest C_0 (fF)	0.185	N.A	N.A
Pitch (µm)	12	12	5
Thermal Conductance G (W/K)	5×10^{-8}	5×10^{-9}	1.5×10^{-8}
Thermal Capacity C (J/K)	26×10^{-12}	80×10^{-12}	3×10^{-12}
Thermal Constant τ_{th} (ms)	0.5	16	0.2
Temperature Sensitivity (/°C)	0.01%	3.6%	0.0092%

At this stage, we have to struggle with a strong signal attenuation due to a capacitive bridge formed by parasitic capacitances from the metallic pads, connections and the input impedance of the final readout electronics board: $V_B \Delta C / (C_0 + C_p)$. The order of magnitude of an expected capacitance variation ΔC is around 10 aF for a $C_0 \sim 200$ aF and $C_p \sim 10$ pF. In this condition, the output signal is divided by a factor ~10^6. This attenuation of the signal can be deleterious for obtaining a high-enough SBR to initiate a self-oscillation within a closed-loop. A way to address this issue can be through the use of a semitone-actuation (at $f_0/2$): $V_{pol} = V_{AC} \cos\left(\frac{2\pi f t}{2}\right) - V_B$. In this case, the electrostatic torque is proportional to $V_{AC}^2/2$, which reduces the coupling between the actuation signal and the output signal. A differential measurement can also be added to further improve the SBR. In this scheme, two identical pixels are used to cancel out the common modes. A more complex approach based on the down-mixing method [45] can be used to get rid of the parasitic capacitances. In particular, the bias voltage is no more constant and is modulated: $V_B = V_{B0} \cos(2\pi f t + \Delta f)$ where $\Delta f \ll f$. A comparison between different readout modes is shown in Table 2. We notice a quite strong improvement of the SBR. However, in the best cases, the signal-to noise-ratio (SNR) was lower than 20 dB, which did not guarantee a functional closed-loop.

Table 2. Preliminary measurement of the signal-to-background ratio (SBR) for different transduction strategies (direct semitone, direct 1f, differential and down-mixed); $V_{B0} = 10$ V, $V_{DC} = 10$ V and $f_0 = 1$ MHz, $\Delta f = 10$ kHz; $V_{AC0} = 4.2$ V for semitone actuation and $V_{AC0} = 0.5$ V for 1f and 2f actuations.

Transduction Method	Voltages		SBR (dB)
-	V_{AC}	V_B	-
1f-actuation	$V_{AC0} \cos(2\pi f t)$	V_{B0}	−33
f/2-actuation	$V_{AC0} \cos\left(\frac{2\pi f t}{2}\right)$	V_{B0}	−13
f/2-actuation/differential mode	$V_{AC0} \cos\left(\frac{2\pi f t}{2}\right)$	V_{B0}	2
f/2-actuation/down-mixing mode	$V_{AC0} \cos\left(\frac{2\pi f t}{2}\right)$	$V_{B0} \cos(2\pi f t + \Delta f)$	22
f-actuation/down-mixing mode	$V_{AC0} \cos(2\pi f t) + V_{DC}$	$V_{B0} \cos(2\pi f t + \Delta f)$	20
2f-actuation/down-mixing mode	$V_{AC0} \cos(4\pi f t + \Delta f)$	$V_{B0} \cos(2\pi f t + \Delta f)$	22

In conclusion, to reach higher SBR (40 dB can be considered as a suitable value for the PLL) a dedicated off-chip buffer was developed to cope with the tiny capacitance variation and it was placed in the close vicinity of the pixels under test. Its schematic is presented in Figure 5. The capacitance variation is read through an intermediate circuit that measures the charge carrier variation, the applied voltage being kept constant. The current is read with a feedback capacitance, C_{C2V}, instead of a resistor to minimize the background, as shown in Figure 5. Thus, the output voltage is proportional to this feedback capacitance as $V_{out} = V_{pol}\frac{\delta C(\theta(f))}{C_{C2V}}$, where $\delta C(\theta_r)$ is the capacitance variation resulting from the motion of the paddle. C_{C2V} must be chosen to be as low as possible to maximize the output signal but should be high enough to avoid unwanted effects due to parasitic capacitances (and $k_B T/C$ noise) as shown in the Equation (6). A 1 pF feedback capacitance in parallel with a 10 GΩ resistor were set to prevent any saturation of the output signal caused by the direct current (DC). C_d and C_p are respectively the sensor capacitance and the total input capacitance of the amplifier, respectively. This electrical scheme corresponds to a first-order filter [44]:

$$V_{out} = V_{pol}\frac{C_d}{C_{C2V}} \left| \frac{1}{\left(1 - j\frac{f_c}{f}\right)\left(1 + j\frac{f}{f_{CO}\frac{C_{C2V}}{C_p}}\right)} \right| \qquad (6)$$

f_{CO} is the high-pass cut-off frequency of the amplifier, and $f_c = \frac{1}{2\pi R_{FB}C_{C2V}}$ the low-pass frequency at -3 dB of the RC filter. The improvement that was brought by the buffer circuit is illustrated in Figure 6. With our buffer, the SBR was kept quite constant, while the SNR strongly improved, growing up to 42 dB (to be compared with 20 dB without the buffer).

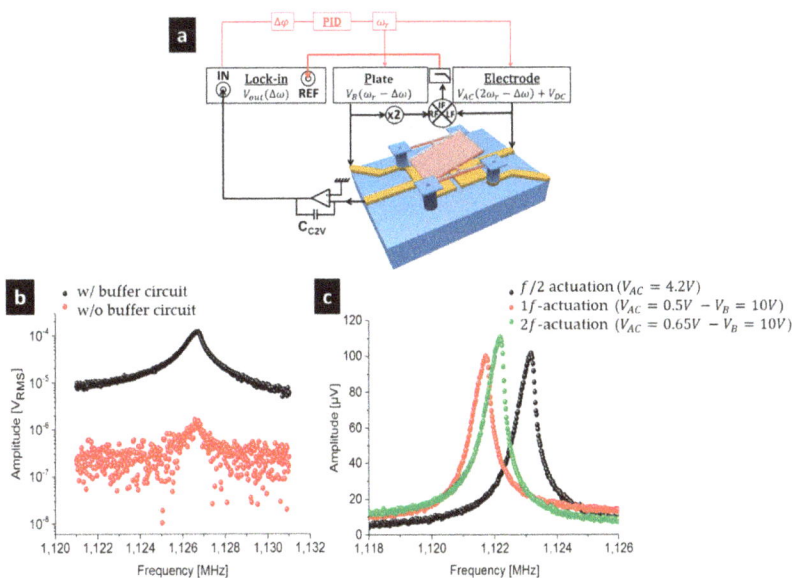

Figure 6. Enhancement provided by the buffer circuit: (**a**) Synopsis showing the set-up to characterize the pixels in a down-mixed readout scheme (in open loop or in a closed-loop: red part); (**b**) Comparison of the output signal between a semitone down-mixing approach with the buffer circuit and without the buffer circuit (the polarization voltages are explained in Table 2); (**c**) Typical output signals for the three down-mixing approaches with the buffer circuit. At resonance, the capacitance variation is close to 10 aF. The SNR is larger than 40 dB for the three cases.

The output voltages at resonance were high and clean enough to embed the electromechanical pixels into a closed loop. At this stage, the pixels of the array of 96 × 96 pixels (see Figure 3a,b) have been tested using an external closed-loop based on a down-mixed phase locked loop (PLL) scheme—only 96 × 96 pixels among the 666 × 520 pixels of the complete array (see Figure 3) could be effectively electrically controlled for use in this first proof of concept. The latter is shown in red in Figure 6a. To do so, the output phase signal $\Delta\varphi$ (demodulated at Δf) was the input signal of a digital proportional–integral–derivative controller (PID). The output f_r corresponded to the frequency applied to the pixel. The PID parameters were set according to the Ziegler-Nichols method [46], hence modifying the bias voltage V_B applied on the pixel. This voltage enabled the control of the effective stiffness [47].

The next section will provide typical electromechanical results and the thermal sensitivity of our electromechanical system. The noise sources of such a system are presented. Based on these measurements, new readout schemes of large pixel arrays are suggested, to achieve a compact CMOS circuit beneath the imager.

3. Results

3.1. Electromechanical Characterizations

First, the electromechanical responses of the pixels of the 96 × 96 array were measured in an open loop. The variations of the resonance amplitude as a function of the voltages V_{B0} and V_{AC0} were verified for the $f/2$ and $2f$ actuation schemes, up to the onset of non-linearity (i.e., $\theta_c \sim 17°$). Figure 7a,b correspond to a $f/2$-actuation showing, as expected, a quadratic variation of the output voltage at resonance with V_{AC0} and a linear variation with V_{B0} respectively. Figure 7c,d correspond to the $2f$-actuation showing a linear variation of the output voltage with V_{AC0} and a quadratic variation with V_{B0}, which also expected.

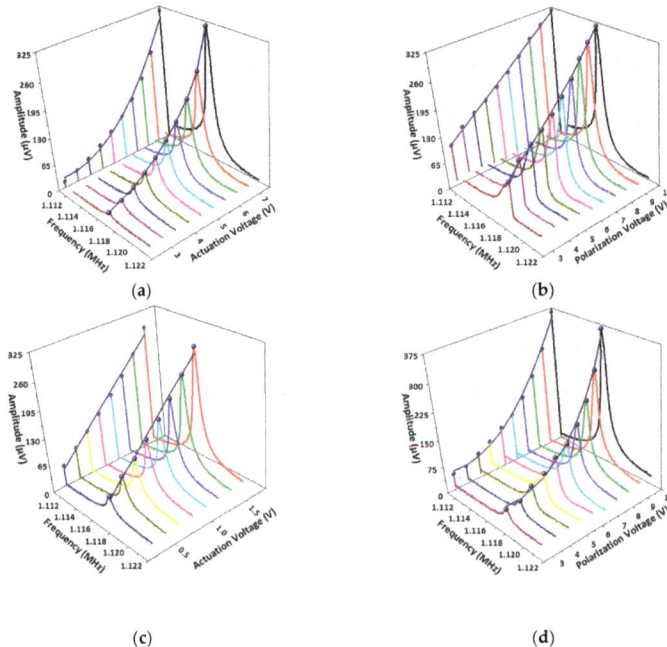

(a)   (b)

(c)   (d)

Figure 7. Electromechanical response when the frequency is swept around the resonance for different actuation schemes: (**a**) Amplitude versus f and V_{AC} for a semitone actuation ($V_B = 10$ V); (**b**) Amplitude versus f and V_B for a semitone actuation ($V_{AC} = 6.5$ V); (**c**) Amplitude versus f and V_{AC} for a $2f$-actuation ($V_B = 10$ V); (**d**) Amplitude versus f and V_B for a $2f$-actuation ($V_{AC} = 1.8$ V).

The main electromechanical features for the torsional mode (f_0, Q and V_{out}, the maximum output voltage corresponding to the onset of nonlinearity of the deflection angle θ_c) were measured on every pixel of a 96 × 96 array with the 2f-actuation scheme:

- $f_0 = [1.05\text{–}1.2\,\text{MHz}]$;
- $Q = [1600\text{–}2500]$;
- $V_{out} = [100\text{–}350\,\mu\text{V}]$

The range of resonance frequencies was coherent, with a fabrication process dispersion of 10% on the torsional rod width (length and thickness variations were negligible). The variations on the quality factor and the maximum voltage were rather more sensitive to the mechanical anchoring of the insulation legs, and the over-etching effect between the edge and the center of the array explained this difference. Spurious flexural motions of the insulation legs (perpendicular to the rods) may have impacted the torsional vibration. At first insight, it increased the damping rate and lowered the effective torsional stiffness. In a more complete approach, the real anchoring features were included into a nonlinear model that was previously presented [47]. In particular, the anchoring was modelled with a stiff spring along the z-axis. We demonstrated that the onset of nonlinearity can be decreased depending on many other factors (the electrostatic torques ...), hence modifying the detector sensitivity. In the worst cases, the thermal sensitivity may be decreased by 10%, compared to the value that is expected with a perfect anchoring.

The frequency dispersion does not impact the closed-loop performance, and it has a tiny impact on the thermal response (see Equation (1)). However, the quality factor and the dynamic range had a larger impact on the noise floor level, as we will see later in Section 3.2. This means that the pixels on the edges of the future imager will be slightly less sensitive compared to the others. To go further on this topic, the next section will address the noise of the readout chain and the thermal performance of such pixels.

3.2. Thermal Characterizations

Let's go back to Equation (1), which gives the thermal response to an incident IR-radiation. The frequency shift is proportional to the temperature coefficient of frequency α_T, and inversely proportional to the thermal conductance G of the material stack of the rods and the insulation legs.

3.2.1. TCF & G

Using the closed loop (Figure 5a), we implemented systematic TCF measurements on typical devices (Figures 3c and 4a–c). In order to carry out a large number of measurements within a reasonable time, the devices were tested on an automatic probe-station that was dedicated to 200 mm wafers. The latter was heated with a hotplate to have a temperature variation between 0° and 20° above the ambient temperature. Beyond this limit, the closed-loop did not track the frequency shift anymore. The measurements were performed with a coupled Peltier-Pt sensor controlled by a Proportional Integral Derivative controller (PID) to obtain the chamber temperature (down to 0.1 °C-accuracy). The statistics are summarized in Table 3.

The TCF of a typical pixel (55.4 ppm/°C) was in good agreement with the theoretical value of 48 ppm found by finite element method simulation (FEM) on our stack. In the case of pixels with thin nano-rods, the axial internal stress was higher and its variation with temperature reinforced the TCF (same sign of variation). In the meantime, the thermal insulation was quite well enhanced. Thus, a global improvement of the thermal response should be expected with this kind of pixel.

Table 3. Temperature coefficient of frequency (TCF) measured on different types of pixels: mean and standard deviation per wafer; thermal conductance of rods & legs G (computed from material properties and geometry measured by SEM).

Pixel Types	$\langle \alpha_T \rangle$ (ppm/°C)	σ_{α_T}	G (W/K)	$\langle \alpha_T \rangle / G$
Typical (Figure 3c)	55.4	14.6	5×10^{-8}	1.11×10^9
Butterfly (Figure 4a)	45.2	3.6	3.10×10^{-8}	1.46×10^9
Typical with Thin Nano-Rod (Figure 4c)	86.2	16.4	1.8×10^{-8}	4.79×10^9

3.2.2. Thermal Response

The thermal response was measured and compared with the theoretical values computed by FEM. Measurements were performed with the readout chain shown in Figure 5a in closed-loop. The device under test (a pixel array) was placed into a vacuum chamber, and a blackbody source (RCN 1200 from HGH Infrared Systems set at 1200 °C) was positioned in front of it. An 8–12 μm-filter was put between our chamber and this source, to control the incident power. The optical bench was aligned thanks to a visible laser. The optical set-up was calibrated using a Fourier transform infrared instrument (FTIR). In particular, the spectral response of the filter according to the spectral luminance of a perfect blackbody at 1200 °C was measured. A photometric computation (knowing the optical apertures and the relative distances between the optical blocks) was used to determine the incident optical power. We considered that the source is a Lambertian black body with a monochromatic luminance, described by Planck's law. The aperture of the source and the chamber window were close enough to neglect the atmosphere absorption in the estimation of incident power ($d = 2.5$ cm). The frequency response of our typical pixel-to-IR incident pulses (17 nW peaks) is presented in Figure 8a). Thermal responses up to $R_f = 1050$ W^{-1} were extracted with the best devices ($f_0 = 1.15$ MHz). Assuming a fill factor $\beta = 0.8$, an efficiency $\eta = 0.8$ in the 8–12 μm window, and considering the measured TCF, $\alpha_T = -76$ ppm/°C, a theoretical thermal response $R_f = 950$ W^{-1} was expected, which was very close to the observed sensitivities. In a second experiment, the incident IR flux on a pixel was changed by varying the distance between the window and the IR-source. The frequency shift was then the measured for optical powers varying from 2.5 to 16 nW. The experimental results and their linear fit are presented in Figure 8b. A thermal response of $R_f = 1350$ W^{-1} was extracted from the slope, considering the resonance frequency mentioned above. Above 8 nW, the relationship between the IR-flux and the thermal frequency shift was no more linear. To increase the incident power, the source was moved closer to the window, which caused it to heat up. This effect lowered its transmittance, resulting in an incorrect estimation of the thermal response (namely, $R_f = 1050$ W^{-1}).

Similar experimental thermal responses were extracted on other pixels of a same array. From one array to another, experimental R_f varied from 700 to 1350 W^{-1}, showing some dispersion attributed to the fabrication process.

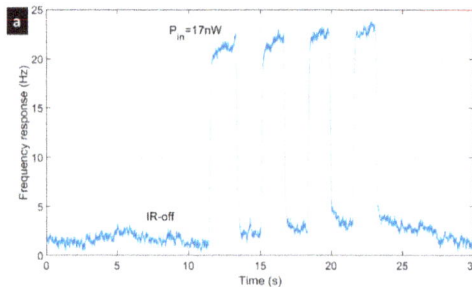

Figure 8. *Cont.*

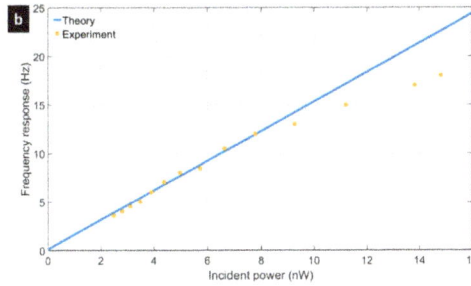

Figure 8. Measurement of the thermal response: (**a**) Resonance frequency shift induced by incident IR pulses (peaks of 17 nW); (**b**) Frequency shift when the incident power is varied from 2 to 16 nW.

3.2.3. Response Time

The response time was measured with a helium neon laser (633 nm) from a commercial Polytec vibrometer (see Figure 9). Since the needed integration time was too short (180 µs) compared to the PLL time constant, this experiment was performed in an open loop scheme (see Figure 6a). We verified that the response time was not limited by the electrical low-pass filter from our measurement set-up, by setting τ = 50 µs. The optical power was set to remain in the linear dynamic range ($\pm 13.5°$ for the typical design). The 10–90% method was used to extract the fall time t_r, and the response time of the first order low-pass filter τ ($\tau = t_r / \ln(9)$). By doing so, we extracted a response time of 430 µs, which was close to the theoretical value computed with the thermal equations (500 µs). The resonant electromechanical pixels could follow quite quick events in the scene. They had a faster response than the current resistive pixels.

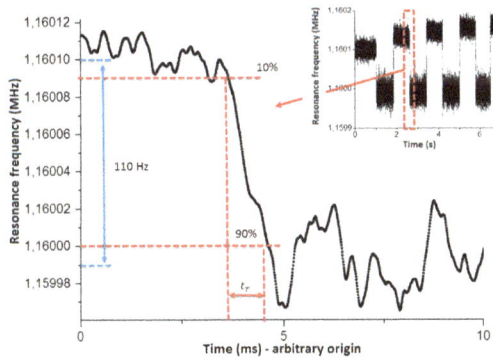

Figure 9. Resonance frequency according to the acquisition time. A 1 mW red laser is focalized onto the pixel under test (typical Figure 3c). Insert: Full data from our measurement of response time. The average frequency jump is estimated as 110 Hz. Then, the response time is extracted from one event fall time (red).

3.3. Noises and Temperature Sensitivity

The performances of the electromechanical pixels were estimated through the noise equivalent power (NEP) or the NETD. The NEP is defined as the incident power on the sensor surface with a SNR of 1. This corresponds to the minimum measurable frequency shift:

$$\text{NEP} = \frac{1}{R_f} \frac{<\delta f^2>^{1/2}}{f_0} = \frac{\sigma_y}{R_f} \qquad (7)$$

$< \delta f^2 >^{1/2}$ is the rms frequency fluctuation for a given bandwidth, and σ_y is the quadratic deviation of the instantaneous relative frequency.

Current "column" or "rolling-shutter" readout schemes should be implemented with our resonator array. The column readout, which is the current approach with CMOS circuit, requires a 60 Hz frame rate, which sets a pixel integration bandwidth at 7 kHz for 190 pixels per column, for instance [48]. However, this readout scheme may induce a lag effect leading to an image distortion when the scene moves faster than the frame rate. A single-pixel readout is a solution to remove this effect. In this case, the integration time corresponded to the full frame rate, i.e., 50 Hz, thereby increasing the SNR. As the capacitive detection did not suffer from self-heating issue, it was possible to use a longer integration time without material degradation, even for small pitches below 12 μm. This is why the NEP of our sensor was estimated for three noise bandwidths, $f_{BW} = 7$ kHz, 50 Hz, and 10 Hz.

Let's get back to a few computational and theoretical considerations to understand and estimate the different noise sources contributing to σ_y. The overall σ_y is the quadratic sum of these noise contributions that are considered as uncorrelated: $\sigma_y = \sqrt{\sum_i \sigma_{yi}^2}$, where i corresponds to: (1) the thermomechanical noise, which is due to the coupling with an ambient thermal bath; (2) the readout electronics' noise, and (3) the phonon noise.

σ_y, whose main origin is the thermomechanical noise, is inversely proportional to the SNR [28,49]:

$$\sigma_y^{Th} = \frac{1}{2Q} \frac{\sqrt{\langle \theta_n^2 \rangle}}{\theta_c} = \frac{1}{2Q} \frac{1}{\text{SNR}} \tag{8}$$

where $\langle \theta_n^2 \rangle$ is the thermomechanical deflection noise. The above expression shows that the linear range and the quality factor must be as high as possible to obtain a stable oscillator. This noise can be estimated through the Parseval-Plancherel theorem: $\langle \theta_n^2 \rangle = \sqrt{\int_{f_0 - \frac{\Delta f}{2}}^{f_0 + \frac{\Delta f}{2}} S_\theta(f) df}$. Using the dissipation-fluctuation theorem, the power spectral density of the thermochemical noise can be written as [50]: $S_\theta(f) = (4\pi k_B T / Q) \kappa f_0^3 / ((f_0^2 - f^2)^2 + (f f_0 / Q))$, with: k_B as the Boltzmann constant, and T as the ambient temperature.

When $f \ll f_0 / Q$ (i.e., when the readout bandwidth is smaller than the mechanical response time; in other words, for $f_{BW} = 50$ Hz), this expression is simplified: $\langle \theta_n^2 \rangle = \frac{2k_B T Q}{\pi \kappa f_0} \sqrt{f_{BW}}$. Interestingly, at a fast integration time (i.e., $f_{BW} = 7$ kHz), $\langle \theta_n^2 \rangle = k_B T / \kappa$, which corresponds to the equipartition energy theorem.

Similarly, σ_y, whose origin is the readout electronics noise, is expressed as:

$$\sigma_y^{elec} = \frac{1}{2Q} \frac{\sqrt{v_n^2}}{V_{Out}} \tag{9}$$

where $\langle v_n^2 \rangle$ is the readout electronics noise generated by the buffer circuit (see Figure 5).

The fundamental source of noise for a thermal conductance that is higher than the radiation conductance [47] should be the phonon noise that results from the random exchange of heat between the sensor and the thermal bath through the mechanical anchors. At thermodynamic equilibrium, the temperature fluctuations due to this fundamental phenomenon can be written as:

$$\sigma_y^{phonon} = \alpha^2 < \overline{\Delta T} >^2 = \begin{cases} \frac{4\alpha^2 k_B T^2}{G_{th}} \Delta f \ if \ f_{BW} < f_{th} \\ \frac{\alpha^2 k_B T^2}{C_{th}} if \ f_{BW} > f_{th} \end{cases} \tag{10}$$

with $f_{th} = 1/4\tau_{th}$ as the thermal cut-off frequency.

The direct phase or frequency fluctuations can be expressed according to a sum of frequency sources with the spectral power density: $S_y(f) = K f^\alpha \ (-4 < \alpha < 2)$.

The orders of magnitudes of the noises and their consequence on the frequency stability and NEP are summarized in Table 4 below, for the two considered bandwidths. We noticed that the readout electronics noise drastically degraded the performance of such a system. In comparison, the NEP of a classical resistive 12 μm-pitch pixel was around 30 pW. This performance could be reached in principle if the electronics noise was minimized. We also mentioned the NEP for another integration time ($f_{BW} = 1$ Hz) that was rather used for gas or mass measurements. If a new readout strategy could be defined with 1s-integration time, the performance would even be better than the current bolometers.

Table 4. Theoretical frequency stability and NEP for the nominal pixel with our electronics and $2f$-down-mixing readout scheme: $R_f = 1050$ /W, $\theta_C = 13.5°$, $V_n = 10$ nV/$\sqrt{\text{Hz}}$, $Q = 2500$ and $V_{out} = 320$ μV.

Noise Sources	$f_{BW} = 50$ Hz			$f_{BW} = 7$ kHz			$f_{BW} = 1$ Hz
	X_n	σ_y	NEP	X_n	σ_y	NEP	NEP
Thermodynamic	3.6×10^{-6} rad	6.3×10^{-9}	6 pW	1.2×10^{-5} rad	2.6×10^{-8}	25 pW	0.85 pW
Electronics	70.7 nV	8.9×10^{-8}	85 pW	836.7 nV	1×10^{-6}	1000 pW	12 pW
Phonon	-	5.8×10^{-9}	5.5 pW		1.8×10^{-9}	17 pW	0.8 pW

Experimental measurements of the frequency stability were achieved to verify our assumption and to think about a specific readout strategy for our electromechanical array. To this end, the frequency stability was measured in closed-loop by estimating the Allan deviation σ_A of the output signal according to the integration time τ. This deviation is a typical tool for the estimation of the stability of oscillators [51]; in particular, to characterize their long-term drift. However, the main types of noises can easily be observed with such a mathematical tool. In particular, a fluctuation of the ambient temperature surrounding the sensor will have an impact that is directly observed on the long term noise. This effect is directly taken into account in the Allan deviation measurements. Besides this point, the sensor has to include blind pixels that provide information on the ambient temperature. The frequency power spectral density was also measured on the same typical pixel.

The Allan deviations measured in the closed loop are presented in Figure 10a for the typical pixel. Between 70 μs and 50 ms σ_A dropped with a $\tau^{1/2}$ slope, showing that white noise was the main contributor in this interval. As shown in Table 4, this trend was mainly attributed to our readout electronics, whose amplitude level was measured at around 40 nv/$\sqrt{\text{Hz}}$. A plateau at 1.5×10^{-7}, appeared between 50 and 200 ms. This $1/f$-noise was well above the noise floor that was normally set by the thermomechanical white noise and phonon noises (close to few 10^{-8} for the two noise sources). Supplementary experiments were achieved to try to understand the origin of this $1/f$-noise. In particular, the Allan deviation was measured for different actuation voltages V_{AC} to increase the maximum output voltage and to improve the SNR. We demonstrated that the plateau is independent of the SNR. We believe that this noise floor is fully inherent in pure frequency fluctuations, whose origin is not clearly identified. Similar noise signatures have been reported as an anomalous phase noise (APN) for flexural nanoresonators [52]. This fundamental noise would only be relevant for small vibrating bodies, which is the case for the nanorods used in the pixel. In the first conclusion, the stability limit of our torsional resonators was set by the APN, and had to be considered as the fundamental limit of our resonant sensors. Even for a 1 s integration time, the NEP would be stuck at around 100 pW. In the discussion section below, we try to figure out this issue.

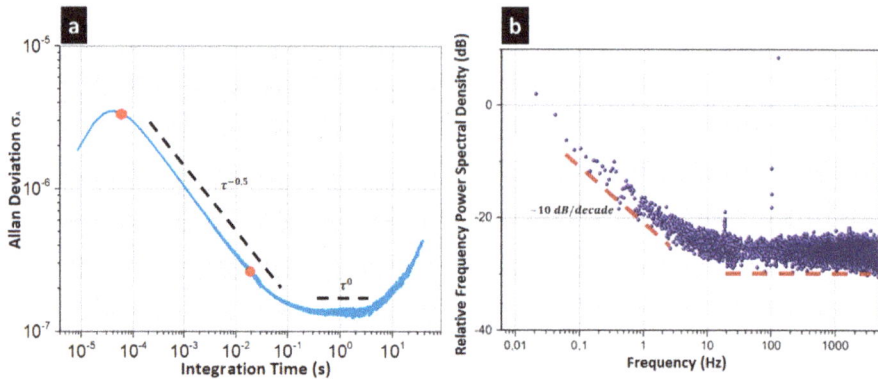

Figure 10. Noise characterization achieved on the typical electromechanical pixel—the amplitude at resonance is set at 320 µV: (**a**) Allan deviation measurement; the red hexagon indicates the frequency deviation for $f_{BW} = 7$ kHz ($\sigma_A = 3.5 \times 10^{-6}$) and the red disk is the one for $f_{BW} = 50$ Hz ($\sigma_A = 3 \times 10^{-7}$); a plateau is reached between 50 and 200 ms integration time ($\sigma_A = 1.5 \times 10^{-7}$); beyond 200 ms, a strong drift effect can be observed; (**b**) spectral power density measurement achieved on the same device. The Allan deviation has a $\tau^{-\frac{1}{2}}$ stop at a short integration time corresponding to the signature of a White amplitude noise. Beyond 50 ms integration time, the Allan deviation presents a plateau, which shows a $1/f$ frequency noise. These two noises can also be distinguished on the power spectral density: the slope of -10 dB/decade corresponds to this plateau.

4. Discussion

First, for the sake of clarity, the NETD, which is basically the lowest temperature variation that is detectable on the scene, is computed for our devices. NETD is directly proportional to the NEP and thus proportional to the frequency stability:

$$\text{NETD} = \frac{4F^2}{\pi A_p \Phi_{\lambda_1 \rightarrow \lambda_2} \left(\frac{\Delta L}{\Delta T} \right)_{300K_{\lambda_1 \rightarrow \lambda_2}}} \times \frac{\sigma_y}{R_f} \tag{11}$$

where F is the optical aperture (usually $F = 1$), A_p is the pixel area, Φ and $(\Delta L/\Delta T)$ are the optical transmission and the luminance variation with a scene temperature of around 300 K, are both evaluated in the $[\lambda_1; \lambda_2]$. range. In the 8–14 µm range, Φ is usually close to one, and $(\Delta L/\Delta T)$ is evaluated as 0.84 W/m^2/sr/K [53].

NETD was computed from the experimental Allan deviations and thermal responses with Equation (11). As a figure of merit, $FOM = NETD \times \tau_{th}$, is usually introduced to evaluate in a glance the quality of a microbolometer technology [54]. Actually this FOM avoids any dependence of the sensor performance to the thermal conductance. These two parameters are shown in Table 5 for our electromechanical components and a current resistive pixel is used as a reference.

Table 5. Comparison between our pixels and a classical resistive bolometer for three integration bandwidths—all electromechanical devices are set at their onset of nonlinearity.

Pixel	τ_{th} (ms)	R_f ($/W_{inc}$)	NEP $f_{BW} = 10$ Hz (nW)	NEP $f_{BW} = 50$ Hz (nW)	NEP $f_{BW} = 7$ kHz (nW)	NETD (K) (FOM—K·ms) $f_{BW} = 10$ Hz	NETD (K) (FOM—K·ms) $f_{BW} = 50$ Hz
Typical	0.5	1050	0.14	0.19	2.4	1.5 (0.75)	2 (1)
Butterfly	0.8	1011	0.47	1.1	12	4.9 (3.96)	11.6 (9.28)
Thin Rod	2.8	3555	1.3	3	35	13.7 (38.3)	-
Resistive Pixel [1]	16	-	-	0.05	-	-	0.05 (4)

A quick insight in Table 5 shows that the NETD of our components cannot compete with the temperature performance of a resistive pixel. Experimental NEP at 50 Hz and 7 kHz bandwidths are close to values obtained through Equations (7), (9), and (11) considering $R_f = 1050$ /W and the experimental value of the electronics noise, 40 nV/$\sqrt{\text{Hz}}$. The NETD of 2 K at 50 Hz and 1.5 K at 100 ms—with a sub-millisecond response time for 8–12 μm incident radiation—were extracted. The usual integration times correspond to the white noise region of our resonator (see Figure 10a) and the NEP can be expressed in terms of a single power density of 30 pW/$\sqrt{\text{Hz}}$. At a long integration time (at 10 Hz), the NETD is set by the APN. In principle, the improvement of the electronics or the thermomechanical noises will not positively influence this limit for long integration times. This analysis tells us that the readout scheme per column at 7 kHz currently used for the resistive bolometer is not suitable for our approach.

From these first conclusions, and if we look at the Equation (11), a few clear avenues of improvement can be proposed:

- Frequency stability and matrix readout strategy: A 50 Hz integration bandwidth requires an improvement of the noise amplitude of our buffer electronics close to the pixel. Lower amplitude noise levels can be reached by using self-oscillating electronics requiring only a few transistors, unlike PLL circuits. Moreover, our electronics was realized close to the pixel but this was not done through an application-specific integrated circuit (ASIC) fabricated underneath the electromechanical pixels. The low-temperature fabrication process presented above has already been used to manufacture resistive bolometer imagers on top of CMOS circuits (ROIC) by post-processing [37,55], and it should be straightforward to reuse this approach in our case. As mentioned in the introductive section of the paper, a co-integration of the readout electronics at the pixel level will reduce the parasitic capacitance down to a few fF, and will decrease the electrical noise down to a theoretical level of 10 nV/$\sqrt{\text{Hz}}$, or even 5 nV/$\sqrt{\text{Hz}}$. This approach makes the down-mixing detection scheme unnecessary, leading to a much simpler measurement chain than the strategy presented here. σ_y will be decreased by a factor of 8 with a self-oscillating IC (a gain of a factor 4 on the absolute noise, and a gain of a factor 2 on the output voltage with a the direct detection (see Equation (9))). Thus, the electronics noise will become lower than the fundamental APN ($\sigma_y = 1.5 \times 10^{-7}$) for a 700 Hz integration bandwidth. This conclusion leads us to suggest a new readout scheme consisting of reading 700/50 = 14 pixels during a 50 Hz frame rate, which allows for a larger area for the co-integrated readout. These two straightforward improvements allow us to obtain a *FOM* that is close to 0.75 for a $f_{BW} = 50$ Hz (global shutter approach), which is an encouraging element.
- Thermal response: At the end, the noise floor level will be set by the APN, whatever the electronics and the readout strategy. An improvement of the signal through the thermal insulation $1/G$ is much trickier in our case. Indeed, this would require long and thin rods/insulations legs, and this would lower the onset of nonlinearity of θ_c (see Equation (8)), leading to a degradation of the SNRs and therefore the frequency stability σ_y.

A simple look at Equation (11) demonstrates that the thermal response R_f has to be increased through the TCF (the temperature coefficient of resistance of a resistive pixel is around 2% whilst the TCF is around 50 ppm). To date, we did not observe a major difference between the experimental TCF values ranging from −35 ppm/°C to −100 ppm/°C. Unfortunately, the highest TCF occurs with soft devices, which were not suitable for IR sensing, as explained above.

We will thus focus the discussion on increasing the TCF of our devices. Some interesting works have already shown a TCF up to 1000 ppm/°C, thereby improving the TCF by a factor of 20 [56,57]. In particular, the first-order phase transition of diverse materials has been used to obtain Young's modulus, which are highly sensitive to temperature [58,59]. Following this line, we are manufacturing similar 12 μm-pitch electromechanical pixels, including VO_2 materials, on top of our pixel. This material was deposited in its amorphous state by reactive deposition (Ion Beam Deposition) and annealed at 400 °C to obtain the crystalline state. The process temperature is kept low enough to be used in a post-process of a CMOS circuit. The Raman characterizations were done to verify the crystallization obtained with this method. The resonators were designed to keep the mechanical features of our current typical pixel (Figure 3c). Nano-indentation measurements were performed on a full layer to extract the Young's modulus of our VO_2 layer (177 GPa for the crystalline state and 80 GPa for the amorphous state). A thickness of 80 nm was then chosen with 1.5 µm long and 300 nm wide torsional rods. In an initial version, both the rods and the plate are covered with the VO_2 layer, and in a second version, the VO_2 layer is only left on the rods. An example of the fabricated devices is shown in Figure 11. The TCF measurements and then the frequency stability are ongoing. We expect an improvement of one order of magnitude in the thermal response (the mechanical features and the thermal insulation being kept constant compared to the standard pixel).

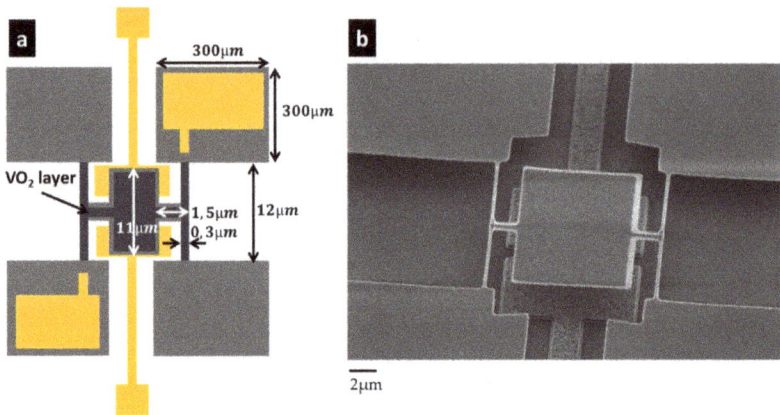

Figure 11. Torsional resonator design: (**a**) Schematics of the design; (**b**) SEM image of a typical pixel with a VO_2 layer on top of both the plate and torsional rods, plus partially on the insulation legs.

5. Conclusions

Our electro-optical measurements show that our current electromechanical resonant pixels cannot compete with the best 12 µm pitch resistive bolometer in terms of NETD. Three major straightforward improvements can be done: (1) the buffer and the electronics readout, including the addressing circuit, has to be included into a ROIC directly beneath the imager as the bolometer. Doing so, the electronics noise and the parasitic capacitance will be negligible, regarding the other sources of noises. The noise level will be set by the APN, which is the fundamental limit of such an approach. This limitation can be overcome by increasing it by a factor of 10- or 100-fold greater than the frequency response. We showed in this paper one of the more promising ways to reach this goal, by integrating phase transition materials on top of the rods. The first realizations demonstrated that we were able to reproduce the same device without thermal features degradations. The optimizations of the pixel are on-going. With both the improvements of the frequency response of a factor of 10, and a pixelwise readout, or at least a readout of 14 in the same frame rate, the NEP would be lower than 20 pW (i.e., NETD < 180 mK). The *FOM* would drop to 0.09, and this value is comparable with the current technology (*FOM* = 0.05). Based on this projection, we believe that the uncooled IR sensors based on the nanomechanical resonators will experience a new interest for small pitches below 12 µm.

Author Contributions: Conceptualization, L.D. and J.-J.Y.; Methodology: L.L. & J.-S.M.; Validation: L.L.; Formal Analysis & Investigation: L.L.; Writing—Original Draft Preparation: L.D.; Writing—Review & Editing: J.A.; Supervision: L.D. and J.-J.Y.; Project Administration: J.-J.Y.; Funding Acquisition: J.-J.Y. & L.D.

Funding: This research was funded by the LETI Carnot Institute's MOTION project.

Acknowledgments: The authors thank P. Imperinetti, M. Sansa and G. Jourdan for their helpful supports with the component fabrication and characterizations.

Conflicts of Interest: The authors declare no conflict of interest.

References

1. Rugar, D.; Budakian, R.; Mamin, H.J.; Chui, B.W. Single spin detection by magnetic resonance force microscopy. *Nature* **2004**, *430*, 329. [CrossRef] [PubMed]
2. Arlett, J.L.; Maloney, J.R.; Gudlewski, B.; Muluneh, M.; Roukes, M.L. Self-sensing micro- and nanocantilevers with attonewton-scale force resolution. *Nano Lett.* **2006**, *6*, 1000–1006. [CrossRef]
3. Sage, E.; Brenac, A.; Alava, T.; Morel, R.; Dupré, C.; Hanay, M.S.; Roukes, L.M.; Duraffourg, L.; Masselon, C.; Hentz, S. Neutral particle mass spectrometry with nanomechanical systems. *Nat. Commun.* **2015**, *6*, 6482. [CrossRef] [PubMed]
4. Hanay, M.S.; Kelber, S.; Naik, A.K.; Chi, D.; Hentz, S.; Bullard, E.C.; Colinet, L.; Duraffourg, E.; Roukes, M.L. Single-protein nanomechanical mass spectrometry in real time. *Nat. Nanotechnol.* **2012**, *7*, 602–608. [CrossRef] [PubMed]
5. Bargatin, I.; Myers, E.B.; Aldridge, J.S.; Marcoux, C.; Brianceau, P.; Duraffourg, L.; Colinet, E.; Hentz, S.; Andreucci, S.; Roukes, M.L. Large-scale integration of nanoelectromechanical systems for gas sensing applications. *Nano Lett.* **2012**, *12*, 1269–1274. [CrossRef] [PubMed]
6. Fahad, H.M.; Shiraki, H.; Amani, M.; Zhang, C.; Hebbar, V.S.; Gao, W.; Ota, H.; Hettick, M.; Kiriya, D.; Chen, Y.-Z.; et al. Room temperature multiplexed gas sensing using chemical-sensitive 3.5-nm-thin silicon transistors. *Sci. Adv.* **2017**, *3*, e1602557. [CrossRef] [PubMed]
7. Ralph, J.E.; King, R.C.; Curran, J.E.; Page, J.S. Miniature quartz resonator thermal detector. In Proceedings of the IEEE Ultrasonics Symposium, San Francisco, CA, USA, 16–18 October 1985; p. 362.
8. Cabuz, C.; Shoji, S.; Fukatsu, K.; Cabuz, E.; Minami, K.; Esashi, M. Fabrication and packaging of a resonant infrared sensor integrated in silicon. *Sens. Actuators A* **1994**, *43*, 92–99. [CrossRef]
9. Zhang, X.C.; Myers, E.B.; Sader, J.E.; Roukes, M.L. Nanomechanical torsional resonators for frequency-shift infrared thermal sensing. *Nano Lett.* **2013**, *13*, 1528–1534. [CrossRef] [PubMed]
10. Yamazaki, T.; Ogawa, S.; Kumagai, S.; Sasaki, M. A novel infrared detector using highly nonlinear twisting vibration. *Sens. Actuators A* **2014**, *212*, 165–172. [CrossRef]
11. Jensen, K.; Kim, K.; Zettl, A. An atomic-resolution nanomechanical mass sensor. *Nat. Nanotechnol.* **2008**, *3*, 533–537. [CrossRef] [PubMed]
12. Chaste, J.; Eichler, A.; Moser, J.; Ceballos, G.; Rurali, R.; Bachtold, A. A nanomechanical mass sensor with yoctogram resolution. *Nat. Nanotechnol.* **2012**, *7*, 301–304. [CrossRef] [PubMed]
13. Ruz, J.J.; Tamayo, J.; Pini, V.; Kosaka, P.M.; Calleja, M. Physics of nanomechanical spectrometry of viruses. *Sci. Rep.* **2014**, *4*, 6051. [CrossRef] [PubMed]
14. Aspelmeyer, M.; Kippenberg, T.J.; Marquardt, F. Cavity optomechanics. *Rev. Mod. Phys.* **2014**, *86*, 1391. [CrossRef]
15. Metcalfe, M. Applications of cavity optomechanics. *Appl. Phys. Rev.* **2014**, *1*, 031105. [CrossRef]
16. Li, B.B.; Bulla, D.; Bilek, J.; Prakash, V.; Forstner, S.; Sheridan, E.; Madsen, L.; Rubinsztein-Dunlop, H.; Foster, S.; Schäfermeier, C.; et al. Ultrasensitive and broadband magnetometry with cavity optomechanics. In Proceedings of the IEEE Conference on Lasers and Electro-Optics (CLEO), San Jose, CA, USA, 14–19 May 2017.
17. Godin, M.; Bryan, A.K.; Burg, T.P.; Babcock, K.; Manalis, S.R. Measuring the mass, density, and size of particles and cells using a suspended microchannel resonator. *Appl. Phys. Lett.* **2007**, *91*, 123121. [CrossRef]
18. Agache, V.; Blanco-Gomez, G.; Baleras, F.; Caillat, P. An embedded microchannel in a MEMS plate resonator for ultrasensitive mass sensing in liquid. *Lab Chip* **2011**, *11*, 2598–2603. [CrossRef] [PubMed]
19. Min, D.H.; Moreland, J. Quantitative measurement of magnetic moments with a torsional resonator: Proposal for an ultralow moment reference material. *J. Appl. Phys.* **2005**, *97*, 10R504. [CrossRef]
20. Antonio, D.; Dolz, M.I.; Pastoriza, H. Micromechanical magnetometer using an all-silicon nonlinear torsional resonator. *Appl. Phys. Lett.* **2009**, *95*, 133505. [CrossRef]

21. Davis, J.P.; Vick, D.; Li, P.; Portillo, S.K.N.; Fraser, A.E.; Burgess, J.A.J.; Fortin, D.C.; Hiebert, W.K.; Freeman, M.R. Nanomechanical torsional resonator torque magnetometry. *J. Appl. Phys.* **2011**, *109*, 07D309. [CrossRef]

22. Perez-Morelo, D.; Pastoriza, H. Torque magnetometry in YBa$_2$Cu$_3$O$_7$-δ single crystals using high sensitive micromechanical torsional oscillator. *J. Phys. Conf. Ser.* **2014**, *568*, 22038. [CrossRef]

23. Degen, C.L.; Poggio, M.; Mamin, H.J.; Rettner, C.T.; Rugar, D. Nanoscale magnetic resonance imaging. *Proc. Natl. Acad. Sci. USA* **2009**, *106*, 1313–1317. [CrossRef] [PubMed]

24. Loh, O.Y.; Espinosa, H.D. Nanoelectromechanical contact switches. *Nat. Nanotechnol.* **2012**, *7*, 283–295. [CrossRef] [PubMed]

25. Wong, A.-C.; Nguyen, C.T.C. Micromechanical mixer-filters ('mixlers'). *J. Microelectromech Syst.* **2004**, *13*, 100–112. [CrossRef]

26. Arlett, J.L.; Myers, E.B.; Roukes, M.L. Comparative advantages of mechanical biosensors. *Nat. Nanotechnol.* **2011**, *6*, 203–215. [CrossRef] [PubMed]

27. Tamayo, J.; Kosaka, P.M.; Ruz, J.J.; San Paulo, Á.; Calleja, M. Biosensors based on nanomechanical systems. *Chem. Soc. Rev.* **2013**, *42*, 1287–1311. [CrossRef] [PubMed]

28. Duraffourg, L.; Arcamone, J. From MEMS to NEMS. In *Nanoelectromechanical Systems*, 1st ed.; Baptist, R., Duraffourg, L., Eds.; ISTE Ltd.: London, UK, 2015; Chapter 1; p. 5. ISBN 978-1-84821-669-3.

29. Mile, E.; Jourdan, G.; Bargatin, I.; Labarthe, S.; Marcoux, C.; Andreucci, P.; Hentz, S.; Kharrat, C.; Colinet, E.; Duraffourg, L. In-plane nanoelectromechanical resonators based on silicon nanowire piezoresistive detection. *Nanotechnology* **2010**, *21*, 165504. [CrossRef] [PubMed]

30. Colinet, E.; Duraffourg, L.; Labarthe, S.; Hentz, S.; Robert, P.; Andreucci, P. Self-oscillation conditions of a resonant nanoelectromechanical mass sensor. *J. Appl. Phys.* **2009**, *105*, 124908. [CrossRef]

31. Sage, E.; Sansa, M.; Fostner, S.; Defoort, M.; Gély, M.; Naik, A.K.; Morel, R.; Duraffourg, L.; Roukes, M.L.; Alava, T.; et al. Single-particle mass spectrometry with arrays of frequency-addressed nanomechanical resonators. *arXiv*, **2017**. [CrossRef]

32. Endoh, T.; Tohyama, S.; Yamazaki, T.; Tanaka, Y.; Okuyama, K.; Kurashina, S.; Miyoshi, M.; Katoh, K.; Yamamoto, T.; Okuda, Y.; et al. Uncooled infrared detector with 12 μm pixel pitch video graphics array. In Proceedings of the SPIE Conference on Defense, Security, and Sensing, Baltimore, MD, USA, 11 June 2013; Volume 8704, p. 87041G.

33. FLIR Boson® | FLIR Systems. Available online: http://www.flir.com/cores/boson/ (accessed on 24 February 2017).

34. TWV640 Thermal Camera Core | BAE Systems | International. Available online: http://www.baesystems.com/en/product/twv640-thermal-camera-core (accessed on 24 February 2017).

35. Jo, Y.; Kwon, I.-W.; Kim, D.S.; Shim, H.B.; Lee, H.C. A self-protecting uncooled microbolometer structure for uncooled microbolometer. In Proceedings of the SPIE Conference on Defense, Security, and Sensing, Orlando, FL, USA, 25–29 April 2011; Volume 8012, p. 80121O.

36. Dorn, D.A.; Herrera, O.; Tesdahl, C.; Shumard, E.; Wang, Y.-W. Impacts and mitigation strategies of sun exposure on uncooled microbolometer image sensors. In Proceedings of the SPIE Conference on Defense, Security, and Sensing, Orlando, FL, USA, 25–29 April 2011; Volume 8012, p. 80123Z.

37. Tissot, J.L.; Durand, A.; Garret, T.; Minassian, C.; Robert, P.; Tinnes, S.; Vilain, M. High performance uncooled amorphous silicon VGA IRFPA with 17 μm pixel-pitch. In Proceedings of the SPIE Conference on Defense, Security, and Sensing, Orlando, FL, USA, 5–9 April 2010; Volume 7660, p. 76600T.

38. Fraenkel, A.; Mizrahi, U.; Bikov, L.; Adin, A.; Malkinson, E.; Giladi, A.; Seter, D.; Kopolovich, Z. VOx-based uncooled microbolometric detectors: Recent developments at SCD. In Proceedings of the SPIE Conference on Defense, Security, and Sensing, Orlando, FL, USA, 17 May 2006; Volume 6206, p. 62061C.

39. Truitt, P.A.; Hertzberg, J.B.; Huang, C.C.; Ekinci, K.L.; Schwab, K.C. Efficient and Sensitive Capacitive Readout of Nanomechanical Resonator Arrays. *Nano Lett.* **2007**, *7*, 120–126. [CrossRef] [PubMed]

40. Younis, M.I. *MEMS Linear and Nonlinear Statics and Dynamics*; Springer: Boston, MA, USA, 2011; Volume 20.

41. Verbridge, S.S.; Parpia, J.M.; Reichenbach, R.B.; Bellan, L.M.; Craighead, H.G. High quality factor resonance at room temperature with nanostrings under high tensile stress. *J. Appl. Phys.* **2006**, *99*, 124304. [CrossRef]

42. Kacem, N.; Hentz, S.; Pinto, D.; Reig, B.; Nguyen, V. Nonlinear dynamics of nanomechanical beam resonators: Improving the performance of NEMS-based sensors. *Nanotechnology* **2009**, *20*, 275501. [CrossRef] [PubMed]

43. Kozinsky, I.; Postma, H.W.C.; Bargatin, I.; Roukes, M.L. Tuning nonlinearity, dynamic range, and frequency of nanomechanical resonators. *Appl. Phys. Lett.* **2006**, *88*, 253101. [CrossRef]

44. Analog Devices, Op Amp Total Output Noise Calculations for Second-Order System. Available online: http://www.analog.com/media/en/training-seminars/tutorials/MT-050.pdf (accessed on 1 June 2016).

45. Bargatin, I.; Myers, E.B.; Arlett, J.; Gudlewski, B.; Roukes, M.L. Sensitive detection of nanomechanical motion using piezoresistive signal downmixing. *Appl. Phys. Lett.* **2005**, *86*, 133109. [CrossRef]

46. Ziegler, J.G.; Nichols, N.B. Optimum Settings for Automatic Controllers. *J. Dyn. Syst. Meas. Control* **1993**, *115*, 220–222. [CrossRef]

47. Laurent, L.; Yon, J.J.; Moulet, J.S.; Imperinetti, P.; Duraffourg, L. Compensation of nonlinear hardening effect in a nanoelectromechanical torsional resonator. *Sensors Actuators A* **2017**, *263*, 326–331. [CrossRef]

48. Wood, R.A. Chapter 3—Monolithic Silicon Microbolometer Arrays. In *Uncooled Infrared Imaging Arrays and Systems*; Semiconductors and Semimetals Series; Kruse, W., Skatrud, D.D., Eds.; Elsevier: New York, NY, USA, 1997; Volume 47, p. 43.

49. Robins, W.P. *Phase Noise in Signal Sources: Theory and Applications*; IET: London, UK, 1984.

50. Cleland, A.N.; Roukes, M.L. Noise processes in nanomechanical resonators. *J. Appl. Phys.* **2002**, *92*, 2758–2769. [CrossRef]

51. Allan, D.W. Time and Frequency (Time-Domain) Characterization, Estimation, and Prediction of Precision Clocks and Oscillators. *IEEE Trans. Ultrason. Ferroelectr. Freq. Control* **1987**, *34*, 647–654. [CrossRef] [PubMed]

52. Sansa, M.; Sage, E.; Bullard, E.C.; Gély, M.; Alava, T.; Colinet, E.; Naik, A.K.; Villanueva, L.G.; Duraffourg, L.; Roukes, M.L.; et al. Frequency fluctuations in silicon nanoresonators. *Nat. Nanotechnol.* **2016**, *11*, 552–558. [CrossRef] [PubMed]

53. Kruse, P.W.; Skatrud, D.D. *Uncooled Infrared Imaging Arrays and Systems*; Academic Press: Cambridge, MA, USA, 1997; Volume 47.

54. Skidmore, G.D.; Han, C.J.; Li, C. Uncooled microbolometers at DRS and elsewhere through 2013. In Proceedings of the SPIE Conference on Image Sensing Technologies: Materials, Devices, Systems, and Applications, Baltimore, MD, USA, 21 May 2014; Volume 9100, p. 910003.

55. Tissot, J.-L.; Mottin, E.; Martin, J.-L.; Yon, J.-J.; Vilain, M. Advanced uncooled infrared focal plane development at CEA/LETI. In Proceedings of the International Conference on Space Optics, Rhodes Island, Greece, 4–8 October 2010; p. 1056902. [CrossRef]

56. Schodowski, S.S. Resonator self-temperature-sensing using a dual-harmonic-mode crystal oscillator. In Proceedings of the 43rd Annual Symposium on Frequency Control, Denver, CO, USA, 31 May–2 June 1989; pp. 2–7. [CrossRef]

57. Li, M.-H.; Chen, C.-Y.; Li, C.-S.; Chin, C.-H.; Li, S.-S. Design and Characterization of a Dual-Mode CMOS-MEMS Resonator for TCF Manipulation. *J. Microelectromech. Syst.* **2015**, *24*, 446–457. [CrossRef]

58. Holsteen, A.; Kim, I.S.; Lauhon, L.J. Extraordinary Dynamic Mechanical Response of Vanadium Dioxide Nanowires around the Insulator to Metal Phase Transition. *Nano Lett.* **2014**, *14*, 1898. [CrossRef] [PubMed]

59. Rúa, A.; Cabrera, R.; Coy, H.; Merced, E.; Sepúlveda, N.; Fernández, F.E. Phase transition behavior in microcantilevers coated with M1-phase VO_2 and M_2-phase VO_2:Cr thin films. *J. Appl. Phys.* **2012**, *111*, 104502. [CrossRef]

micromachines

MDPI

Article

Micro Magnetic Field Sensors Manufactured Using a Standard 0.18-μm CMOS Process

Yen-Nan Lin and Ching-Liang Dai *

Department of Mechanical Engineering, National Chung Hsing University, Taichung 402, Taiwan;
g102061073@mail.nchu.edu.tw
* Correspondence: cldai@dragon.nchu.edu.tw; Tel.: +886-4-2284-0433

Received: 29 June 2018; Accepted: 6 August 2018; Published: 7 August 2018

Abstract: Micro magnetic field (MMF) sensors developed employing complementary metal oxide semiconductor (CMOS) technology are investigated. The MMF sensors, which are a three-axis sensing type, include a magnetotransistor and four Hall elements. The magnetotransistor is utilized to detect the magnetic field (MF) in the x-axis and y-axis, and four Hall elements are used to sense MF in the z-axis. In addition to emitter, bases and collectors, additional collectors are added to the magnetotransistor. The additional collectors enhance bias current and carrier number, so that the sensor sensitivity is enlarged. The MMF sensor fabrication is easy because it does not require post-CMOS processing. Experiments depict that the MMF sensor sensitivity is 0.69 V/T in the x-axis MF and its sensitivity is 0.55 V/T in the y-axis MF.

Keywords: micro sensor; Hall effect; magnetic field; magnetotransistor

1. Introduction

Magnetic field (MF) sensors, which are important components, are applied in industrial apparatuses, automation equipment, cable-stayed bridges, electrical devices and portable electronic instruments [1–4]. Traditional MF sensors [5,6], which were not manufactured by microfabrication, were not only large volume, but also high cost. The advantages of micro magnetic field (MMF) sensors are small volume and low cost. Micro-electro-mechanical-system (MEMS) technology could be utilized to fabricate micro sensors [7–11]. Several MMF sensors were manufactured using MEMS technology. For instance, Mian [12] developed resonant MMF sensors fabricated by the surface micromachining process. The sensor structure contained microbeams and a membrane, the material of which was a stack of double polysilicon layers. Based on the Lorentz force principle, the MMF caused a capacitance change upon sensing an MF. The Lorentz force resonant MEMS magnetic field sensors were proposed by Park [13]. The MMF sensors had a micromirror actuated by the Lorentz force that was generated using a sinusoidal current and an incident MF. The rotation angle of the micromirror was recorded using an optical measurement. Dennis [14] used a CMOS process to manufacture resonant MMF sensors. The sensor was fabricated using the stacked layers of the CMOS process, and a post-CMOS processing with reactive ion etch (RIE) dry etching was adopted, releasing the device structure. The sensor shuttle was excited using the Lorentz force and external MF, and the resonance amplitude was detected by an optical instrument. These resonant MMF sensors [12–14] required movable suspension structures, so sacrificial layer technology was used to release the suspension structures. For example, after completion of the CMOS process, the resonant MMF sensors, proposed by Dennis [14], used an RIE dry etching post-processing to obtain the suspension structures of the devices. In this work, we design a magnetotransistor/Hall element MMF sensor without a suspension structure using a CMOS process, so the sensor does not need post-CMOS processing. Therefore, the sensor fabrication in this work is easier than that of these sensors [12–14].

A one-axis magnetotransistor MMF sensor, presented by Tseng [15], was fabricated by a standard 0.18-μm CMOS process of Taiwan Semiconductor Manufacturing Company (TSMC). The MMF sensor had a sensitivity of 354 mV/T. Furthermore, Tseng [16] adopted the same method to design a three-axis magnetotransistor MMF sensor that was also made using a standard 0.18-μm CMOS process of TSMC. The MMF sensor had a sensitivity of 6.5 mV/T in the x-axis and a sensitivity of 0.4 mV/T in the y-axis. A two-dimensional Hall MMF sensor with a lateral magnetotransistor and magnetoresistor, developed by Yu [17], was produced using a standard 0.35-μm CMOS process. The sensitivity of the MMF sensor was 0.385 V/(A·T) at a bias current of 100 mA. Sung [18] proposed a two-dimensional Hall MMF sensor manufactured utilizing a standard 0.35-μm CMOS process. The MMF sensor was composed of a bulk magnetotransistor, a vertical magnetoresistor and a vertical magnetotransistor. The sensitivity of the sensor was 1.92 V/(A·T) at a bias current of 20 mA. With the same design method, Sung [19] developed a three-dimensional Hall MMF sensor with a bandgap reference and readout circuit made using a standard 0.18-μm CMOS process. The MMF sensor contained one-dimensional lateral Hall sensor and a two-dimensional vertical Hall sensor. The MMF sensor had a sensitivity of 5943 V/(A·T) at a bias current of 6.25 mA in the x- and y-axis MF and a sensitivity of 14,790 V/(A·T) at a bias current of 6.25 mA in the z-axis MF. The Hall MMF sensors, proposed by Xu [20], were fabricated by the 0.18-μm high voltage (HV) CMOS process for sensing low MF. The sensors consisted of a Hall plate with a switching cross-shape. Zhao [21] utilized a CMOS process to make nano-polysilicon transistor MMF sensors. A nano-polysilicon/single silicon junction was adopted as a sensing layer. The nano-polysilicon transistors were fabricated on silicon substrate with high resistivity. The two-dimensional MMF sensors, develop by Yang [22], included four magnetic transistors. The MMF sensors were manufactured on a high resistivity silicon wafer using microfabrication technology, and they were packaged on printed circuit boards. The sensor sensitivity in the x-axis was 366 mV/T, and its sensitivity in the y-axis was 365 mV/T, respectively. These micro sensors [15,17,20–22] manufactured using CMOS technology were one-axis and two-axis MMF sensors. Therefore, three-axis MMF sensors in this work are fabricated using a standard 0.18-μm CMOS process of TSMC, and the sensitivity of the sensors is higher than that of Tseng [16].

Various MEMS actuators and sensors, which are manufactured utilizing a CMOS process, are called CMOS-MEMS devices [23–27]. We adopt CMOS-MEMS technology to develop three-axis MMF sensors. MMF sensors are composed of a magnetotransistor and four Hall elements. The magnetotransistor is designed to detect MF in the x-axis and y-axis. Four Hall elements are designed to sense MF in the z-axis. These CMOS-MEMS magnetic field sensors [28–30] needed a post-CMOS processing [31] to form suspension structures. The fabrication of the MMF sensors in this study is consistent with the CMOS process and does not need post-CMOS processing.

2. Structure of MMF Sensor

Figure 1a demonstrates the MMF sensor structure, where E denotes the emitter, B_1, B_2, B_3 and B_4 are the bases, C_1, C_2, C_3 and C_4 are the collectors, AC_1, AC_2, AC_3 and AC_4 are the additional collectors and H_1, H_2, H_3, H_4, H_5, H_6, H_7 and H_8 are the electrodes of the Hall elements.

The MMF sensor includes a magnetotransistor and four Hall elements. The magnetotransistor is employed to detect MF in the x-axis and y-axis, and the four hall elements are utilized to sense MF in the z-axis. In addition to the emitter, bases and collectors, the additional collectors are introduced into the magnetotransistor. The additional collectors can increase bias current, so that the emitter induces more electron carriers. Shallow trench isolation (STI) oxide is used to confine the current direction, so that leakage current reduces. Figure 2 illustrates the equivalent circuit for the magnetotransistor, where R represents the resister, V_{C1}, V_{C2}, V_{C3} and V_{C4} are the bias voltage of the collectors, V_{AC1}, V_{AC2}, V_{AC3} and V_{AC4} are the bias voltage of the additional collectors, V_{B1}, V_{B2}, V_{B3}, and V_{B4} are the bias voltage of the bases, $V_{out\text{-}AC1}$, $V_{out\text{-}AC2}$, $V_{out\text{-}AC3}$ and $V_{out\text{-}AC4}$ are the output voltages and the symbol (4) denotes the corresponding circuit repeated four times.

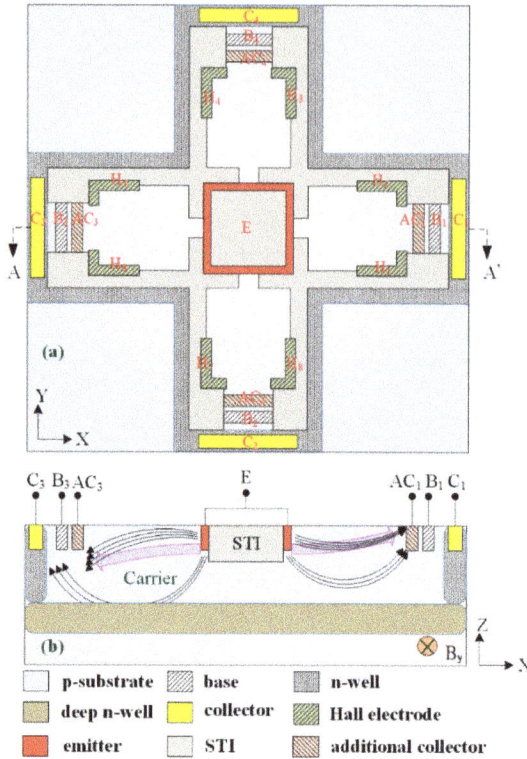

Figure 1. Micro magnetic field (MMF) sensor: (**a**) schematic structure; (**b**) cross-sectional view along line AA'. STI, shallow trench isolation; E, emitter.

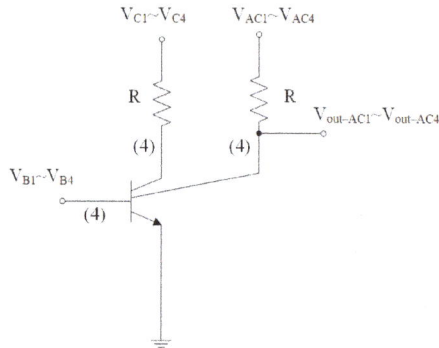

Figure 2. Equivalent circuit for the magnetotransistor. V, bias voltage; C, collector; R, resistor; AC, additional collector; B, base.

Figure 1b demonstrates a cross-sectional view of the MMF sensor. The sensing mechanism of the magnetotransistor is explained as follows. As shown in Figure 1b, when the bias voltages are applied to the collectors, the bases and the additional collectors, carriers produce a movement from the emitter to the additional collectors AC_1/AC_3, the bases B_1/B_3 and the collectors C_1/C_3. Given a magnetic field in the y-axis, carriers (on the right in Figure 1b) are bent upward because of the action of the Lorentz force. The carriers have difficulty passing across the additional collector AC_1 to the base B_1 and collector C_1. Most of carriers move to the additional collector AC_1, resulting in the current increment of the additional collector AC_1. At the same time, carriers (on the left in Figure 1b) are bent downward owing to the action of the Lorentz force. Most of the carriers move across the additional collector AC_3 to the base B_3 and collector C_3, leading to the current decrement of the additional collector AC_3. Therefore, this action produces a voltage difference between the additional collectors AC_1 and AC_3 in the y-direction MF. As shown in Figure 2, the additional collectors AC_1 and AC_3 respectively connect to a resistor R, so the voltage difference of the additional collectors AC_1 and AC_3 is obtained by $V_{out-AC1} - V_{out-AC3}$, which is the sensor output voltage in the y-axis MF. Similarly, when the bias voltages are applied to the collectors, the bases and the additional collectors, carriers produce a movement from the emitter to the additional collectors AC_2/AC_4, the bases B_2/B_4 and the collectors C_2/C_4. Given a magnetic field in the x-axis, carriers that move to the additional collector AC_2 are bent downward by Lorentz force, leading to the current decrement of the additional collector AC_2. Additionally, carriers that move to the additional collector AC_4 are bent upward by the Lorentz force, resulting in the current increment of the additional collector AC_4. The current between both additional collectors AC_2 and AC_4 generates an imbalance, so the additional collectors AC_2 and AC_4 produce a voltage difference. As shown in Figure 2, the voltage difference of the additional collectors AC_2 and AC_4 can be obtained by $V_{out-AC4} - V_{out-AC2}$, which is the sensor output voltage in the x-axis MF.

The MMF sensor has four Hall elements used to detect z-direction MF. Figure 3 presents the carriers' path in the MMF sensor under z-direction MF. As shown in Figure 3, when the bias voltages are applied to the collectors, bases and additional collectors, carriers cause a movement from the emitter to the additional collectors AC_1, AC_2, AC_3 and AC_4. Given an MF in the z-axis, carriers are bent toward the Hall electrodes H_1, H_3, H_5 and H_7 by the Lorentz force. The current causes an imbalance between the electrodes H_1 and H_2, leading to the generation of a Hall voltage between the electrodes H_1 and H_2. Similarly, the Hall voltages between the electrodes H_3/H_4, H_5/H_6 and H_7/H_8, respectively, are generated in z-direction MF. The Hall voltages in series are the sensor output voltages in z-axis MF.

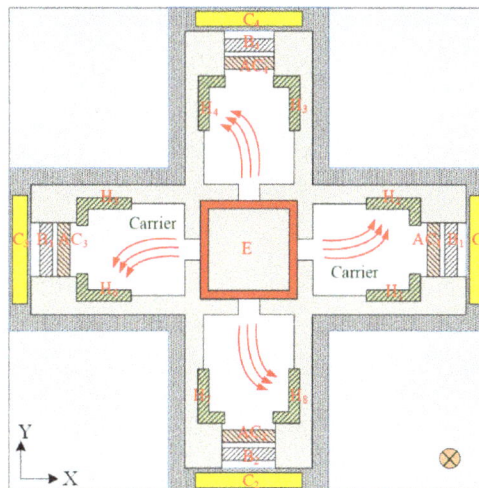

Figure 3. Carrier path in the MMF sensor under z-direction MF. H, Hall electrode, E, emitter.

The Sentaurus TCAD, which is a finite element method software, was utilized to simulate the MMF sensor performance. According to the structure in Figure 1a, the model of the MMF sensor was constructed. To save computation time, one-quarter of the MMF model was established because the

MMF sensor structure was symmetric. Then the method of Delaunay triangulation was employed to mesh the MMF model. The approach of the Poisson electron hole was used to compute the coupling effect of MF and the electrical field, and the method of Bank/Rose was utilized to solve the carrier density distribution of the MMF sensor.

In this simulation, bias voltages of 1.5 V, 4.5 V and 4.5 V were supplied to bases, collectors and additional collectors, respectively. A magnetic field of 250 mT was given in the y-axis. Figure 4 shows the simulated carrier density distribution of the MMF sensor under the y-direction magnetic field. Figure 4a illustrates the carrier density distribution of the MMF sensor without a magnetic field. Figure 4b reveals the carrier density distribution of the MMF sensor with a magnetic field of 250 mT in the y-axis. By the comparison of the simulated results in Figure 4a,b, the current density of the path from the emitter to the additional collector increases.

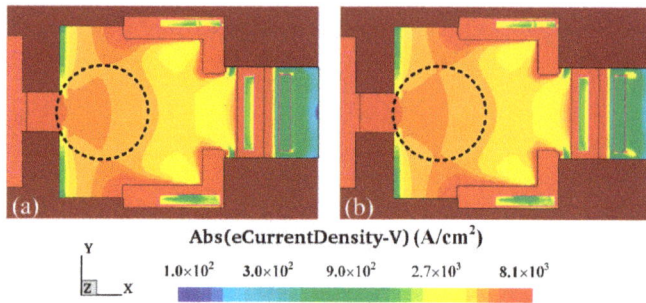

Figure 4. Carrier density distribution of the MMF sensor under the y-direction magnetic field.

The carrier density distribution of the MMF sensor in the z-direction magnetic field was computed with the same simulation approach. In this computation, bias voltages of 1.5 V, 4.5 V and 4.5 V were supplied to bases, collectors and additional collectors, respectively. A magnetic field of 250 mT was applied in the z-axis. Figure 5 shows the simulated carrier density distribution of the MMF sensor under the z-direction magnetic field. Figure 5a demonstrates the carrier density distribution of the MMF sensor without a magnetic field. Figure 5b presents the surface carrier density distribution of the MMF sensor with a magnetic field of 200 mT in the z-axis. As illustrated in Figure 5a,b, carriers are bent toward the top Hall electrode.

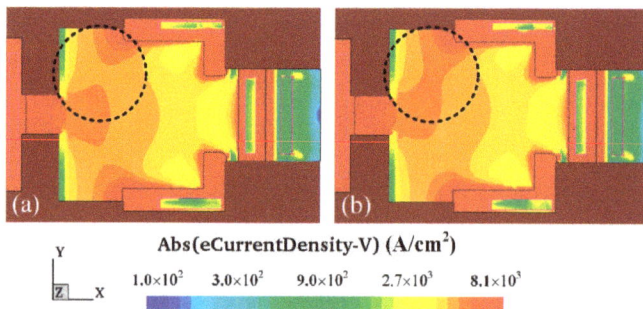

Figure 5. Carrier density distribution of the MMF sensor under the z-direction magnetic field.

3. Fabrication of MMF Sensor

The layout of the MMF sensor was designed in accordance with the structure in Figure 1a. The TSMC used a 0.18-μm CMOS process to manufacture the MMF sensor according to the MMF sensor layout. Figure 6 demonstrates the cross-sectional structure of the MMF sensor after completion of the CMOS process. As demonstrated in Figure 6, the MMF sensor was fabricated on p-type substrate. The MMF sensor consisted of an emitter, four collectors, four bases, four additional collectors and eight Hall electrodes. The emitter was n-type silicon doping phosphorus. The collectors and the additional collectors were n-type silicon doping phosphorus, and the bases were p-type silicon doping boron. Hall electrodes were n-type silicon doping boron. The deep n-well layer, which was a buried layer, was connected to n-well layer to confine the current downward moving range and to reduce leakage current. The STI oxide, which surrounded the emitter edge to confine the current moving direction, would reduce leakage current. An image of the MMF sensor is presented in Figure 7a. Figure 7b shows the magnified picture of the sensor with a scale bar. Fabrications of the other MEMS MMF sensors [12–14] were more complicated than that of the MMF sensor because the MMF sensor did not require any post-processing. The MMF sensor chip was wire-bonded and packaged on a frame. Figure 7c shows the MMF sensor picture after packaging.

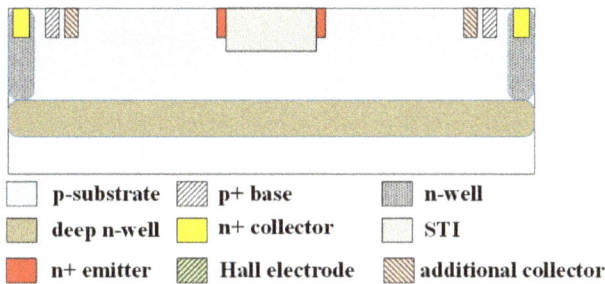

Figure 6. Cross-sectional view of the MMF sensor after the CMOS process.

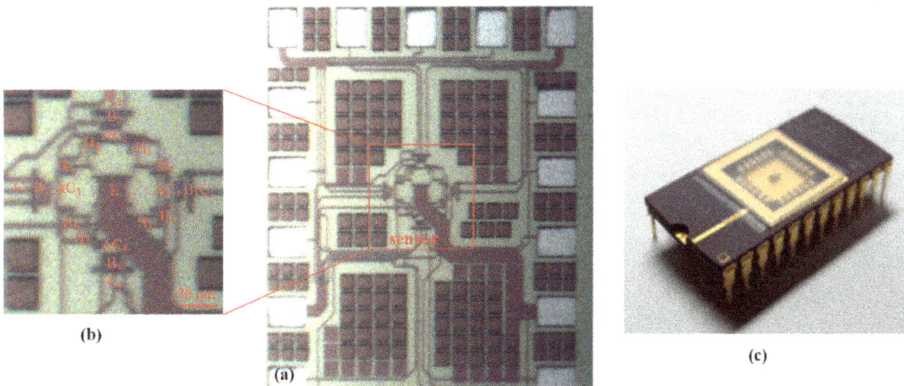

Figure 7. Picture of the MMF sensor chip.

4. Results

A magnetic testing system was employed to measure the MMF sensor performance. Figure 8 demonstrates a magnetic testing system [15], and the system includes a Gauss-meter (GM08-1029, Hirst, Falmouth, U.K.), an MF generator (developed by our lab), a power supply (GPC-3030DQ, Gwinstek, New Taipei City, Taiwan) and a digital multimeter (34405A, Agilent, Santa Clara, CA, USA). The magnetic generator was employed to generate an MF to test the MMF sensor. The power supply was used to provide power to the MF generator. The Gauss-meter was used to test the magnetic magnitude excited by the MF generator. The digital multimeter was utilized to record the MMF sensor output voltage.

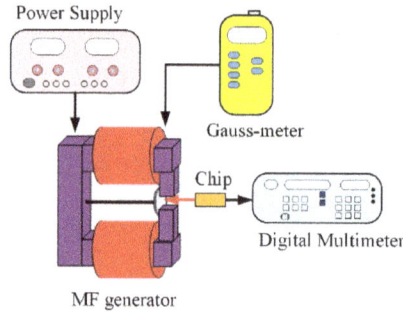

Figure 8. Experiment setup for the MMF sensor.

The MMF sensor was composed of a magnetotransistor and four Hall elements. The magnetotransistor was used to detect MF in the x- and y-directions, and the Hall elements were utilized to measure the magnetic field in the z-direction. First, the MMF sensor performance in the x-direction MF was tested. The MMF sensor was set in the magnetic testing system. An MF range of −220–220 mT generated by the MF generator was supplied to the MMF sensor, and the MF magnitude was calibrated using the Gauss-meter. Bias voltages were applied to the bases, collectors and additional collectors. An MF in the y-direction was applied to the MMF sensor. The digital multimeter measured the voltage difference of the additional collectors AC_1/AC_3 in the MMF sensor. Figure 9 depicts the measured output voltage of the MMF sensor in the y-direction MF. When $V_B = 1$ V, $V_C = 0.6$ V and $V_{AC} = 0.6$ V, the sensor was insensitive to MF. The sensor became more sensitive to MF at $V_B = 1.25$ V, $V_C = 2.04$ V and $V_{AC} = 2.04$ V, and its output voltage changed from −43.7 mV at −220 mT to 38.6 mV at 220 mT. When $V_B = 1.5$ V, $V_C = 3.38$ V and $V_{AC} = 3.38$ V, the sensor output voltage obviously increased under different MF. When $V_B = 1.75$ V, $V_C = 4.58$ V and $V_{AC} = 4.58$ V, the sensor output voltage varied from −120 mV at −220 mT to 119 mV at 220 mT, and the method of least squares was adopted to evaluate the linear regression of the curve. The evaluation obtained that the regression line had a slope of 0.55 V/T and a standard deviation of 5.4 mV. Thereby, the MMF sensor sensitivity in the y-direction MF was 0.55 V/T at bias voltage $V_B = 1.75$ V, $V_C = 4.58$ V and $V_{AC} = 4.58$ V.

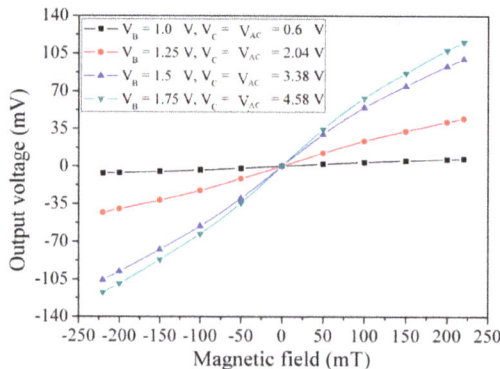

Figure 9. Measured output voltage in the y-direction MF.

With the same testing approach, the sensing performance of the MMF sensor in the x-direction MF was measured. A MF in the x-direction was applied to The MMF sensor. The digital multimeter recorded the voltage difference of the additional collectors AC_2/AC_4 in the MMF sensor. Figure 10 demonstrates the measured output voltage of the MMF sensor in the x-direction MF. The sensor was insensitive to MF at $V_B = 1$ V, $V_C = 0.6$ V and $V_{AC} = 0.6$ V. The sensor was more sensitive to MF at $V_B = 1.25$ V, $V_C = 2.04$ V and $V_{AC} = 2.04$ V, and it output voltage varied from -46.8 mV at -220 mT to 39 mV at 220 mT. When $V_B = 1.5$ V, $V_C = 3.38$ V and $V_{AC} = 3.38$ V, the output voltage enlarged under different MF. When $V_B = 1.75$ V, $V_C = 4.58$ V and $V_{AC} = 4.58$ V, the output voltage changed from -162 mV at -220 mT to 140 mV at 220 mT, and the method of least squares was used to calculate the linear regression of the curve. The calculation obtained that the regression line had a slope of 0.69 V/T and a standard deviation of 12 mV. Thereby, the MMF sensor sensitivity in the x-direction MF was 0.69 V/T at bias voltage $V_B = 1.75$ V, $V_C = 4.58$ V and $V_{AC} = 4.58$ V.

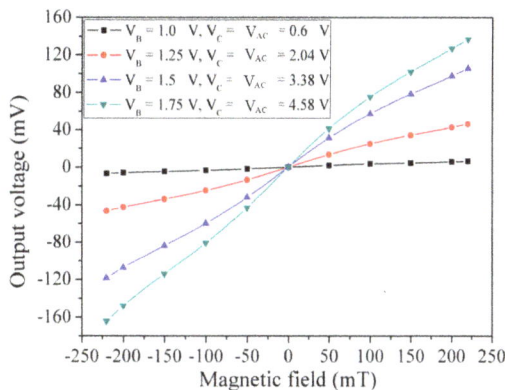

Figure 10. Measured output voltage in the x-direction MF.

The sensing performance of the MMF sensor in the z-direction MF was tested. An MF in the z-direction was applied to the MMF sensor. The digital multimeter measured the output voltage of the Hall electrodes in the MMF sensor. The Hall voltage of the MMF sensor in the z-direction MF was recorded. Figure 11 presents the measured output voltage of the MMF sensor in the z-direction MF. The MMF sensor was also insensitive to MF at $V_B = 1$ V, $V_C = 0.6$ V and $V_{AC} = 0.6$ V. When bias voltage V_B, V and V_{AC} increased, the sensor output voltage became large. When $V_B = 1.75$ V, $V_C = 4.58$ V and $V_{AC} = 4.58$ V, the MMF sensor output voltage varied from -20.5 mV at -220 mT to 20 mV at 220 mT, the curve slope of which was about 0.09 V/T. Thereby, the MMF sensor sensitivity in the z-direction MF was 0.09 V/T at bias voltage $V_B = 1.75$ V, $V_C = 4.58$ V and $V_{AC} = 4.58$ V.

The characteristics of the MMF sensor in the x- and y-direction MF should be the same owing to the MMF sensor being a symmetric structure. Actually, as shown in Figure 9, the measured sensitivity of the MMF sensor in the y-direction MF was 0.55 V/T. There is little difference in the sensitivity of the MMF sensor in the x- and y-direction MF. The reason is due to packaging and fabrication deviation. As shown in Figure 11, the curves were linear, because the sensing mechanism of the MMF sensor in the z-axis MF was based on the Hall elements. The Hall voltage, which was the output voltage of the sensor in the z-axis MF, was proportional to the magnetic field according to the sensing principle of the Hall element [20]. On the other hand, the sensing mechanism of the MMF sensor in the x- and y-axis MF was based on the magnetotransistor. The carrier current density (Figure 3) of the magnetotransistor depended on the magnetic field, and the I-V (current-voltage) characteristic of magnetotransistor was nonlinear [15], so that the output voltage versus magnetic field (Figures 9 and 10) of the MMF sensor in the x- and y-axis MF was nonlinear.

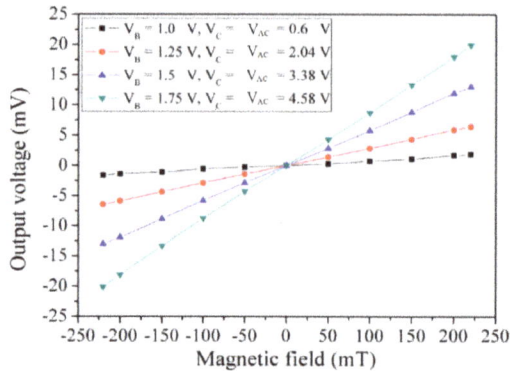

Figure 11. Measured output voltage in the z-direction MF.

Table 1 shows a list of sensitivity for various MMF sensors fabricated by CMOS technology. The MMF sensors presented by Tseng [15], Xu [20] and Zhao [21] were one-axis MF sensing, and the sensors proposed by Yu [17], Sung [18] and Yang [22] were two-axis MF sensing. As depicted in Table 1, the sensitivity of the MMF sensor in this work along the x- and y-axis MF exceeds that of Tseng [16], Yu [17], Sung [18] and Yang [22]. The sensitivity of the sensor presented by Zhao [21] along the z-axis MF is higher than that of this work.

Table 1. Sensitivity of various MMF sensors.

MMF Sensors	Sensitivity (mV/T)		
	x-axis	y-axis	z-axis
Tseng [15]	354	-	-
Tseng [16]	6.5	6.5	0.4
Yu [17]	38.5	38.5	-
Sung [18]	38.4	38.4	-
Xu [20]	-	-	31
Zhao [21]	-	-	264
Yang [22]	366	365	-
This work	690	550	90

5. Conclusions

Micro magnetic field sensors were developed employing the CMOS-MEMS technique. The MMF sensors included a magnetotransistor and four Hall elements, where the magnetotransistor could sense MF in the x-axis and y-axis and four Hall elements could detect MF in the z-axis. Additional collectors were added to the magnetotransistor, so that bias current increased and the emitter induced more electron carriers, resulting in enhancing the sensor sensitivity. Manufacturing of the MMF sensors was simple as they did not require post-CMOS processing. Experiments showed that the MMF sensor sensitivity was 0.69 V/T in the x-direction MF at bias voltage V_B = 1.75 V, V_C = 4.58 V and V_{AC} = 4.58 V, and its sensitivity was 0.55 V/T in the y-direction MF at bias voltage V_B = 1.75 V, V_C = 4.58 V and V_{AC} = 4.58 V. The sensitivity was 0.09 V/T in the z-direction MF at bias voltage V_B = 1.75 V, V_C = 4.58 V and V_{AC} = 4.58 V.

Author Contributions: Conceptualization, C.-L.D. Data curation, Y.-N.L. Investigation, Y.-N.L. Methodology, Y.-N.L. Project administration, C.-L.D. Supervision, C.-L.D. Writing review and editing, C.-L.D.

Acknowledgments: The authors would like to thank National Center for High-performance Computing (NCHC) for chip simulation; National Chip Implementation Center (CIC) for chip fabrication and the Ministry of Science and Technology (MOST) of the Republic of China for financially supporting this research under Contract Nos. MOST 105-2221-E-005-037-MY3 and MOST 106-2221-E-005-050-MY2.

Conflicts of Interest: The authors declare no conflict of interest.

References

1. Bougas, L.; Wilzewski, A.; Dumeige, Y.; Antypas, D.; Wu, T.; Wickenbrock, A.; Bourgeois, E.; Nesladek, M.; Clevenson, H.; Braje, D.; et al. On the possibility of miniature diamond-based magnetometers using waveguide geometries. *Micromachines* **2018**, *9*, 276. [CrossRef]
2. Guo, L.; Wang, C.; Zhi, S.; Feng, Z.; Lei, C.; Zhou, Y. Wide linearity range and highly sensitive MEMS-based micro-fluxgate sensor with double-layer magnetic core made of Fe–Co–B amorphous alloy. *Micromachines* **2017**, *8*, 352. [CrossRef]
3. Lara-Castro, M.; Herrera-May, A.L.; Juarez-Aguirre, R.; Lopez-Huerta, F.; Ceron-Alvarez, C.A.; Cortes-Mestizo, I.E.; Morales-Gonzalez, E.A.; Vazquez-Leal, H.; Dominguez-Nicolas, S.M. Portable signal conditioning system of a MEMS magnetic field sensor for industrial applications. *Microsyst. Technol.* **2017**, *23*, 215–223. [CrossRef]
4. Valadeiro, J.; Cardoso, S.; Macedo, R.; Guedes, A.; Gaspar, J.; Freitas, P.P. Hybrid integration of magnetoresistive sensors with MEMS as a strategy to detect ultra-Low magnetic fields. *Micromachines* **2016**, *7*, 88. [CrossRef]
5. Can, H.; Topal, U. Design of ring core fluxgate magnetometer as attitude control sensor for low and high orbit satellites. *J. Supercond. Nov. Magn.* **2015**, *28*, 1093–1096. [CrossRef]
6. Zhang, X.; Liu, B.; Zhang, H.; Wu, J.; Song, B.; Wang, C. A magnetic field sensor based on a dual S-tapered multimode fiber interferometer. *Meas. Sci. Technol.* **2018**, *29*, 075103. [CrossRef]
7. Yang, M.Z.; Dai, C.L.; Lin, W.Y. Fabrication and characterization of polyaniline/PVA humidity microsensors. *Sensors* **2011**, *11*, 8143–8151. [CrossRef] [PubMed]
8. Zhao, Y.; Zhao, Y.; Ge, X. The development of a triaxial cutting force sensor based on a MEMS strain gauge. *Micromachines* **2018**, *9*, 30. [CrossRef]
9. Huang, J.Q.; Li, B.; Chen, W. A CMOS MEMS humidity sensor enhanced by a capacitive coupling structure. *Micromachines* **2016**, *7*, 74. [CrossRef]
10. Hu, Y.C.; Dai, C.L.; Hsu, C.C. Titanium dioxide nanoparticle humidity microsensors integrated with circuitry on-a-chip. *Sensors* **2014**, *14*, 4177–4188. [CrossRef] [PubMed]
11. Liao, W.Z.; Dai, C.L.; Yang, M.Z. Micro ethanol sensors with a heater fabricated using the commercial 0.18 μm CMOS process. *Sensors* **2013**, *13*, 12760–12770. [CrossRef] [PubMed]
12. Mian, M.U.; Dennis, J.O.; Khir, M.H.M.; Ahmed, M.G.A.; Rabih, A.A.S.; Tang, T.B. Experimental analysis of out-of-plane Lorentz force actuated magnetic field sensor. *IEICE Electron. Express.* **2017**, *14*, 20161257. [CrossRef]
13. Park, B.; Li, M.T.; Liyanage, S.; Shafai, C. Lorentz force based resonant MEMS magnetic-field sensor with optical readout. *Sens. Actuators A Phys.* **2016**, *241*, 12–18. [CrossRef]
14. Dennis, J.O.; Ahmad, F.; Khir, M.H.B.M.; Hamid, N.H.B. Optical characterization of Lorentz force basedCMOS-MEMS magnetic field sensor. *Sensors* **2015**, *15*, 18256–18269. [CrossRef] [PubMed]
15. Tseng, J.Z.; Wu, C.C.; Dai, C.L. Modeling and manufacturing of a micromachined magnetic sensor using the CMOS process without any post-process. *Sensors* **2014**, *14*, 6722–6733. [CrossRef] [PubMed]
16. Tseng, J.Z.; Shih, P.J.; Hsu, C.C.; Dai, C.L. A three-axis magnetic field microsensor fabricated utilizing a CMOS process. *Appl. Sci.* **2017**, *7*, 1289. [CrossRef]
17. Yu, C.P.; Sung, G.M. Two-dimensional folded CMOS Hall device with interacting lateral magnetotransistor and magnetoresistor. *Sens. Actuators A Phys.* **2012**, *182*, 6–15. [CrossRef]
18. Sung, G.M.; Yu, C.P. 2-D differential folded vertical Hall device fabricated on a p-type substrate using CMOS technology. *IEEE Sens. J.* **2013**, *13*, 2253–2262. [CrossRef]

19. Sung, G.M.; Gunnam, L.C.; Wang, H.K.; Lin, W.S. Three-dimensional CMOS differential folded Hall sensor with bandgap reference and readout circuit. *IEEE Sens. J.* **2018**, *18*, 517–527. [CrossRef]

20. Xu, Y.; Pan, H.B.; He, S.Z.; Li, L. A highly sensitive CMOS digital Hall sensor for low magnetic field applications. *Sensors* **2012**, *12*, 2162–2174. [CrossRef] [PubMed]

21. Zhao, X.F.; Wen, D.Z.; Zhuang, C.C.; Liu, G.; Wang, Z.Q. High sensitivity magnetic field sensors based on nano-polysilicon thin-film transistors. *Chin. Phys. Lett.* **2012**, *29*, 118501. [CrossRef]

22. Yang, X.; Zhao, X.; Bai, Y.; Lv, M.; Wen, D. Two-dimensional magnetic field sensor based on silicon magnetic sensitive transistors with differential structure. *Micromachines* **2017**, *8*, 95. [CrossRef]

23. Dai, C.L.; Peng, H.J.; Liu, M.C.; Wu, C.C.; Hsu, H.M.; Yang, L.J. A micromachined microwave switch fabricated by the complementary metal-oxide semiconductor post-process of etching silicon dioxide. *Jpn. J. Appl. Phys.* **2005**, *44*, 6804–6809. [CrossRef]

24. Qu, H. CMOS MEMS fabrication technologies and devices. *Micromachines* **2016**, *7*, 14. [CrossRef]

25. Dai, C.L.; Chen, H.L.; Chang, P.Z. Fabrication of a micromachined optical modulator using the CMOS process. *J. Micromech. Microeng.* **2001**, *11*, 612–615. [CrossRef]

26. Huang, J.Q.; Li, F.; Zhao, M.; Wang, K. A surface micromachined CMOS MEMS humidity sensor. *Micromach.* **2015**, *6*, 1569–1576. [CrossRef]

27. Dai, C.L.; Hsu, H.M.; Tsai, M.C.; Hsieh, M.M.; Chang, M.W. Modeling and fabrication of a microelectromechanical microwave switch. *Microelectron. J.* **2007**, *38*, 519–524. [CrossRef]

28. Beroulle, V.; Bertrand, Y.; Latorre, L.; Nouet, P. Monolithic piezoresistive CMOS magnetic field sensors. *Sens. Actuators A Phys.* **2003**, *103*, 23–32. [CrossRef]

29. Hsieh, C.H.; Dai, C.L.; Yang, M.Z. Fabrication and characterization of CMOS-MEMS magnetic microsensors. *Sensors* **2013**, *13*, 14728–14739. [CrossRef] [PubMed]

30. Lu, C.C.; Liu, Y.T.; Jhao, F.Y.; Jeng, J.T. Responsivity and noise of a wire-bonded CMOS micro-fluxgate sensor. *Sens. Actuators A Phys.* **2012**, *179*, 39–43. [CrossRef]

31. Dai, C.L.; Chiou, J.H.; Lu, M.S.C. A maskless post-CMOS bulk micromachining process and its application. *J. Micromech. Microeng.* **2005**, *15*, 2366–2371. [CrossRef]

micromachines

MDPI

Article

AFM-Based Characterization Method of Capacitive MEMS Pressure Sensors for Cardiological Applications

Jose Angel Miguel, Yolanda Lechuga * and Mar Martinez

Group of Microelectronics Engineering, Department of Electronics Technology, Systems Engineering and Automation, University of Cantabria, Santander 39005, Spain; jamd@teisa.unican.es (J.A.M.); martinez@teisa.unican.es (M.M.)
* Correspondence: yolanda@teisa.unican.es; Tel.: +34-942-201-863

Received: 31 May 2018; Accepted: 3 July 2018; Published: 6 July 2018

Abstract: Current CMOS-micro-electro-mechanical systems (MEMS) fabrication technologies permit cardiological implantable devices with sensing capabilities, such as the iStents, to be developed in such a way that MEMS sensors can be monolithically integrated together with a powering/transmitting CMOS circuitry. This system on chip fabrication allows the devices to meet the crucial requirements of accuracy, reliability, low-power, and reduced size that any life-sustaining medical application imposes. In this regard, the characterization of stand-alone prototype sensors in an efficient but affordable way to verify sensor performance and to better recognize further areas of improvement is highly advisable. This work proposes a novel characterization method based on an atomic force microscope (AFM) in contact mode that permits to calculate the maximum deflection of the flexible top plate of a capacitive MEMS pressure sensor without coating, under a concentrated load applied to its center. The experimental measurements obtained with this method have allowed to verify the bending behavior of the sensor as predicted by simulation of analytical and finite element (FE) models. This validation process has been carried out on two sensor prototypes with circular and square geometries that were designed using a computer-aided design tool specially-developed for capacitive MEMS pressure sensors.

Keywords: micro-electro-mechanical systems (MEMS) sensors; MEMS modelling; capacitive pressure sensor; MEMS characterization; atomic force microscope; stent

---

1. Introduction

Cardiovascular diseases are the predominant cause of mortality worldwide [1,2]. The 2013 Global Burden of Disease study estimated that they were responsible for more than double the deaths than that caused by cancer. In the European Union alone, they accounted for almost 40% of all deaths in 2013, and ischemic heart diseases (IHD) alone were responsible for more than 35% of deaths. These conditions are caused by the accumulation of fatty deposits lining the inner wall of a coronary artery, restricting blood flow to the heart.

Patients diagnosed with IHD are commonly subjected to a surgical procedure called percutaneous coronary intervention (PCI), in which the regular blood-flow in a clogged vessel is usually restored and maintained by the implantation of a biocompatible mesh tube or Stent. Nevertheless, neo-intimal tissue growth inside the stent (in-Stent restenosis, ISR) stands out as its major drawback, jeopardizing patients' life and forcing, in many cases, the repetition of the procedure. Current tracking methods for ISR are expensive and time-consuming as they require complex equipment, specialized medical staff, and even patient's hospitalization. Thus, the proposal of intelligent stents (iStent) endowed with blood-flow and/or pressure sensing capabilities represents a potential economic solution that, nonetheless, must be reliable, efficient, compact, low-power, and less expensive than its counterparts to be considered as an actual alternative.

In this sense, the arising of an affordable fabrication technology of micro-electro-mechanical systems (MEMS), which combine an integrated circuit (IC) with mechanical parts, permits these stents with tracking capabilities to meet the aforementioned requirements.

Among the different options for MEMS pressure sensors [3,4], this work is based on a capacitive approach in which sensors are conceived as parallel-plate capacitors with a fixed and a flexible plate that bends with increasing pressure. The IC will measure and transmit the equivalent capacitance of the sensor that will reflect the decrease in the dielectric gap by deflection of the upper plate and, thus the applied pressure.

However, before undertaking the task of developing a monolithic heterogeneous IC that includes the electronic circuitry and the MEMS sensor on the same substrate, it is advisable to characterize stand-alone prototype sensors in an efficient but affordable way to better recognize further areas of improvement. On the one hand, these prototypes are to be fabricated using MEMS technology that allows for the development of several sensor types at a low cost. On the other hand, a fast characterization method should have direct access to the mechanical structure of the sensor in order to directly measure the deflection of the upper plate and, thus to characterize its detection capability and sensitivity.

The simplest approach to MEMS pressure sensor's experimental characterization implies the design and fabrication of a measurements setup, in order to expose the device under test (DUT) to an accurately controlled pressure [5]. However, this characterization methodology is only suitable for fully sealed MEMS sensors, with their inner gap isolated from the outer environment. For the case of biosensors similar to the ones described in this paper, a biocompatible coating is required to both seal the structure and guarantee its reliable operation once implanted. However, a characterization of the uncoated DUT provides significant advantages: First, the evaluation of the sensor sensitivity loss and offset change as a result of the coating [5]. Second, the acquisition of performance data from the bulk sensor highlights the need for further design adjustments, prior to its monolithic integration. Third, the possibility of double-checking the sensor response (before and after coating) helps to improve its reliability, being a critical aspect for the development of implantable electronic devices. Finally, non-requiring pressure-controlled environments for testing setups provides a significant cost-saving solution for the sensor characterization problem [5].

Hence, the performance parameters of non-coated MEMS pressure biosensors cannot be collected from pressure-based measurements, so another method based on an alternative excitation signal must be used. It must be highlighted that the sensor response to this substitute stimulus has to be formerly modelled analytically and/or numerically, in order to allow the validation of the collected experimental data. In this sense, a novel approach for the characterization of uncoated MEMS pressure biosensors, based on an atomic force microscope operating in contact mode, is proposed in this paper. The atomic force microscope (AFM) is a relatively common laboratory instrument, and this methodology takes advantage of its highly sensitive optical lever detection system and ability to position a probe with nanometer precision, in order to apply a known concentrated load directly on the center of the upper and flexible plate of the sensor. By this way, the deflection of the sensor can be calculated according to the concept shown in Figure 1.

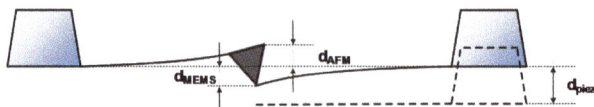

Figure 1. Concept of sensor characterization by the atomic force microscope (AFM) operation in contact mode (not to scale). MEMS—micro-electro-mechanical systems.

In this work two prototypes of capacitive MEMS pressure sensors, with circular and square shapes, have been fabricated using PolyMUMPS technology. This MEMS fabrication technology has been selected for being mature and generalist, in the sense that it can accommodate different MEMS structures and also allows sensor prototyping without coating.

The bending of the AFM probe can be calculated by Hook's law:

$$F = k_{AFM} \cdot \Delta z \tag{1}$$

where F is the force applied by the AFM probe, k_{AFM} is its spring constant, and Δz is the relative displacement of the probe from the equilibrium position where the applied force is zero.

Assuming that once the equilibrium has been reached, the forces acting on the AFM probe and the sensor under test are the same. Thus, it can be stated that $k_{AFM} d_{AFM} = k_{MEMS} d_{MEMS}$, where d_{AFM} and d_{MEMS} are the AFM and sensor deflections, respectively. Considering an AFM piezo vertical displacement d_{piezo} during the measurement required to move the sample position, it is noticeable that d_{piezo} combines both the AFM probe and MEMS sensor displacements, as shown in Figure 1. Thus, the resulting sensor displacement d_{MEMS} can be calculated as follows [6–9]:

$$d_{MEMS} = d_{piezo} - d_{AFM} \tag{2}$$

In Patil et al. [6], an AFM in contact mode is used for the characterization and calibration of a piezoresistive pressure sensor designed for tactile sensing applications. In this paper, the probe-tip was modified by attaching a spherical soda-lime glass particle to its end so as to increase the contact area and simulate a uniform pressure application.

The goal of Alici et al. [7] is to characterize the stiffness of microfabricated cantilevers by utilizing an AFM to develop a static deflection measurement method. More specifically, the AFM is used to apply a known load at the end of a polymer microactuator so as to determine the spring constant from the resulting displacement and a reference calibration method. In this sense, the paper emphasizes the need of a previous calibration step to determine the spring constant of the AFM cantilever.

On the other hand, Rollier et al. [8] take advantage of the AFM features to obtain measurements of force and resonant frequency so that the value of the tensile residual stress of silicon nitride membranes could be extracted. Thus, the objective of this works lies in developing a non-destructive method that permits to improve the low stress silicon nitride deposition process and to optimize released membrane fabrication, not to characterize the behavior of the membrane itself as part of a sensing device.

To summarize, and as stated in Pustan et al. [9], AFMs can be utilized to evaluate the mechanical properties of micro/nanoscale structures and nanomaterials used in MEMS and NanoElectroMechanical Systems (NEMS). More specifically, this work concludes that the dependency between an acting force and the sample deflection is determined by the AFM static mode, and from that data, the stiffness of the microcantilever can be computed. Moreover, the modulus of elasticity of the material is derived by nanoidentation. Besides, AFM-based measurements have been reported to be of great use to characterize pressure sensors, for example, of the piezoresistive type [6]. However, no literature has been found that treated the problem of both analytically and Finite Element (FE) modelling the deflection versus force behavior of non-coated capacitive MEMS pressure biosensors, together with their experimental characterizing using an AFM in contact mode, used to apply a force that permits to estimate the corresponding deflection of the top plate and, thus its sensitivity.

The complete design and characterization process of the MEMS capacitive sensor explained in this work also includes an extensive modelling stage, in which analytical and finite element models have been developed, and are described in Section 2 of this document. This section also includes a description of how a computer-aided design tool specifically developed for capacitive MEMS pressure sensors would guide the designer through the main steps of the process, from the specifications and technology, to the FE model and Cadence layout of the sensor, ready to be sent to the manufacturer. Finally, Section 3 presents the experimental measurements obtained with the AFM compared with the simulation data provided by the analytical and FE models. The results are further discussed in the final section to evaluate the efficiency of the AFM-based characterization method, as well as to propose areas of potential improvement.

2. Materials and Methods

2.1. Capacitive MEMS Pressure Sensor Modelling

The simplest implementation of a pressure monitoring-based iStent, as show in Figure 2a, can be built by attaching one or several MEMS capacitive pressure sensors to the longitudinal ends of a commercially available stent. The device functionality is based on the blood pressure detection performed by the sensors, reflecting on proportional changes in their equivalent capacitance [3,4]. Thus, MEMS sensors act as pressure-dependent capacitances, which, attached to the coil-like stent structure, form an LC-tank whose resonant frequency is modulated by the pressure inside the vessel. An external handheld device is required to perform the wireless communication with the iStent, via inductive coupling techniques, when placed close enough to the implant location.

The topology of a capacitive MEMS pressure sensor, as described in Figure 2b, comprises a fully clamped suspended top plate with thickness t_m, separated a distance of t_g from a backplate fixed to the substrate. As can be noticed, this topology resembles the traditional parallel-plate capacitor build, so its nominal capacitance is ruled by Equation (3):

$$C_0 = \varepsilon_r \varepsilon_0 \frac{A}{t_g} \qquad (3)$$

where ε_r is the relative permittivity of the medium between the plates, ε_0 is the dielectric permittivity of vacuum, and A and t_g are the overlapping area and gap distance between the plates, respectively.

Once a sufficient load p is applied to the sensor, the suspended plate is forced to bend towards the backplate, in such a way that their separation is reduced and the equivalent capacitance C_S is increased. Hence, the resulting load-dependent capacitance can be analytically modeled by Equation (4), where parameter $w(x, y, p)$ refers to the local top-plate deflection.

$$C_S = \int \int_A \frac{\varepsilon_r \varepsilon_0 dx dy}{t_g - w(x, y, p)} \qquad (4)$$

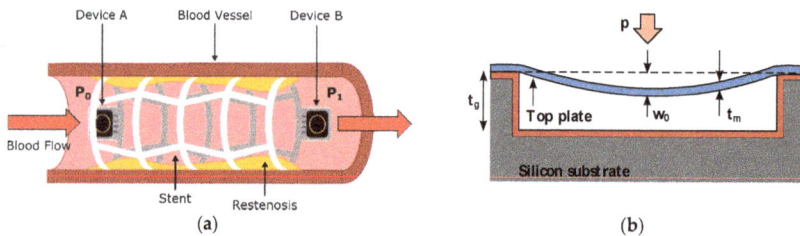

Figure 2. Building parts of a passive pressure sensing iStent. (**a**) Lateral view of the iStent implanted inside a blood vessel; (**b**) cross-sectional view of a MEMS capacitive pressure sensor.

2.1.1. Square Sensor Modelling

As extensively reported in Mechanics books and scientific papers, the governing differential equation for the deflection of a thin plate in cartesian coordinates can be expressed as follows [10,11]:

$$\frac{\partial^4 w}{\partial x^4} + 2 \frac{\partial^4 w}{\partial x^2 \partial y^2} + \frac{\partial^4 w}{\partial y^4} = \frac{p}{D} \qquad (5)$$

where $w(x, y)$ is the deflection of the square plate at any place, p is a distributed load applied to the upper surface of the plate, and D is referred as the flexural rigidity of the plate and can be defined as follows:

$$D = \frac{Et_m^3}{12(1 - v^2)} \tag{6}$$

with E and v being the modulus of elasticity and Poisson's ratio of the plate material, respectively.

In the case of a fully-clamped square plate with side lengths $b = a$ and thickness t_m, such as the one showed in Figure 3a, the following set of constraints describe the plate bending behavior and can be used to solve the differential equation noted in (5).

$$w(x = \pm \frac{a}{2}, y) = 0 \tag{7}$$

$$w(x, y = \pm \frac{b}{2}) = 0 \tag{8}$$

$$\frac{\partial w}{\partial x}(x = \pm \frac{a}{2}, y) = 0 \tag{9}$$

$$\frac{\partial w}{\partial x}(x, y = \pm \frac{b}{2}) = 0 \tag{10}$$

Once a single concentrated load is applied to the center of the square plate (Figure 3b), its deflection equation can be calculated by combining the solutions for three independent problems [10–12]. First, w_1 is the bending solution for a simply supported rectangular plate under a concentrated load located at its center, and can be expressed as follows:

$$w_1(x, y) = \frac{pa^2}{2\pi^3 D} \sum_{m=1,3,5,\ldots} \frac{1}{m^3} \cos \frac{m\pi x}{a} [(\tanh a_m - \frac{a_m}{\cosh^2 a_m}) \cos \frac{m\pi y}{a} - \sinh \frac{m\pi y}{a}$$
$$-\tanh a_m \frac{m\pi y}{a} \sinh \frac{m\pi y}{a} + \frac{m\pi y}{a} \cosh \frac{m\pi y}{a}] \tag{11}$$

where the geometry parameters a_m and β_m, defined as $a_m = m\pi b / 2a$ and $\beta_m = m\pi a / 2b$, can be reduced to $m\pi/2$ for the case of a square plate with sides $a = b$.

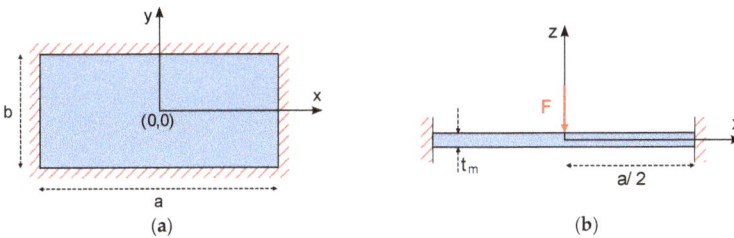

Figure 3. Schematic representation of a rectangular plate under a central concentrated load. (**a**) Top view of the plate; (**b**) cross-sectional view of the plate.

Additionally, w_2 and w_3 are the bending solutions for a simply supported plate with distributed bending moments applied along the edges $y = \pm b/2$ and $x = \pm a/2$, respectively. The applied edge moments are calculated to guarantee a slope at the boundaries equal to zero, as imposed by constraints (9) and (10).

Hence, each aforementioned solution can be defined as follows:

$$w_2(x, y) = -\frac{a^2}{2\pi^2 D} \sum_{m=1,3,5,\ldots} A_m \frac{(-1)^{\frac{m-1}{2}}}{m^2 \cosh a_m} \cos \frac{m\pi x}{a} [\frac{m\pi y}{a} \sinh \frac{m\pi y}{a}$$
$$-a_m \tanh a_m \cosh \frac{m\pi y}{a}] \tag{12}$$

and

$$w_3(x,y) = -\frac{b^2}{2\pi^2 D} \sum_{m=1,3,5,...} B_m \frac{(-1)^{\frac{m-1}{2}}}{m^2 \cosh \beta_m} \cos \frac{m\pi y}{b} \left[\frac{m\pi x}{b} \sinh \frac{m\pi x}{b} \right.$$
$$\left. -\beta_m \tanh \beta_m \cosh \frac{m\pi x}{b} \right] \tag{13}$$

The coefficients A_m and B_m can be determined from the fully-clamped plate constraints (9) and (10), or the condition of cancelling the slope at the plate boundaries. Hence, the most significant A_m and B_m values for a square plate, given in Table 1, can be numerically calculated by successive approximations.

To conclude, the combination of Equations (11)–(13) leads to the final bending solution:

$$w(x,y) = w_1(x,y) + w_2(x,y) + w_3(x,y) \tag{14}$$

Table 1. Values of A_m and B_m parameters for a square plate [12].

m	A_m	B_m
1	$-0.1025 \cdot p$	$-0.1025 \cdot p$
3	$0.0263 \cdot p$	$0.0263 \cdot p$
5	$0.0042 \cdot p$	$0.0042 \cdot p$
7	$0.0015 \cdot p$	$0.0015 \cdot p$
9	$0.00055 \cdot p$	$0.00055 \cdot p$
11	$0.00021 \cdot p$	$0.00021 \cdot p$
13	$0.00006 \cdot p$	$0.000006 \cdot p$

2.1.2. Circular Sensor Modelling

Similar to the case of a square plate, the analytical bending solution for a fully-clamped circular thin plate under a concentrated central force has been widely studied in the literature [10,11]. Hence, differential Equation (5) determining plate deflection can be rearranged in polar coordinates [10,11], and easily applied to a circular plate, such as the one included in Figure 4.

$$\nabla_r^4 w \equiv \left(\frac{\partial^2}{\partial r^2} + \frac{1}{r} \frac{\partial}{\partial r} + \frac{1}{r^2} \frac{\partial^2}{\partial \theta^2} \right) \left(\frac{\partial^2 w}{\partial r^2} + \frac{1}{r} \frac{\partial w}{\partial r} + \frac{1}{r^2} \frac{\partial^2 w}{\partial \theta^2} \right) = \frac{p}{D} \tag{15}$$

with $w(r, \theta)$ being the bending of the circular plate in polar coordinates, and p the distributed load applied to its upper surface. Moreover, the boundary constraints for the plate center and perimeter $(r = a)$ can be expressed as follows:

$$w(r = a, \theta) = 0 \tag{16}$$

and

$$\frac{\partial w}{\partial r}(r = a, \theta) = 0 \tag{17}$$

Unlike the previous case, the bending solution for the fully clamped circular sensor can be obtained in a purely analytical way. By differentiating the bending solution for a simply supported plate and forcing the slope to nullify at the boundary, as imposed by (17), it is possible to calculate the bending moments along the plate edges. The final solution can be calculated by adding the deflection produced by the moments along the edges to the initial simply supported bending equation, obtaining the following expression:

$$w(r, \theta) = \frac{pr^2}{8\pi D} \ln \frac{r}{a} + \frac{p}{16\pi D} (a^2 - r^2) \tag{18}$$

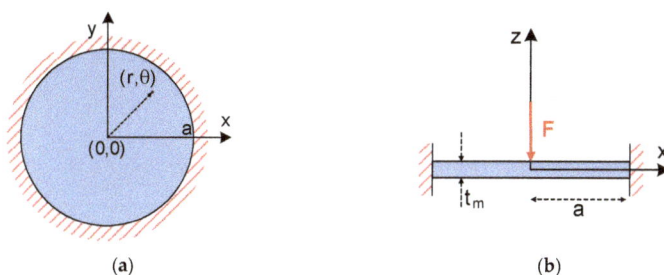

Figure 4. Schematic representation of a circular plate under a central concentrated load. (**a**) Top view of the plate; (**b**) cross-sectional view of the plate.

2.1.3. Modelling Results Comparison

In order to characterize the accuracy of the bending solutions presented in the previous subsections, the responses provided by analytical equations have been compared with the ones obtained from equivalent finite elements (FE) models; so that the full scale error (FSE) between both modelling approaches could be calculated.

Two capacitive polysilicon-based ($E = 169$ GPa; $v = 0.22$) MEMS pressure sensors, circular and square-shaped, have been designed. A self-developed computed aided design (CAD) tool described in Section 2.2 has been used for this purpose. An initial design constraint has been imposed to both sensors, forcing their dimensioning to reach an equal nominal capacitance of $C_0 = 0.6$ pF.

As can be seen in Tables 2 and 3, two different FE models with variable complexity have been developed in ANSYS, in order to perform displacement versus force simulations and compare the resulting bending data with the response anticipated by the analytical expressions (14) and (18). A first FE model, referred as "simple", consists of a flat square or circular plate fully clamped along its edges. This model requires low computational time in ANSYS to achieve a complete bending characterization of the sensor, because of its relatively low complexity. On the other hand, the second FE model, named "complex", presents the exact same topology as the prototype sensor fabricated in PolyMUMPS technology. In this case, hole and dimple elements have been added because of the requirement imposed by the manufacturer for diaphragms larger than 30 μm × 30 μm [13]. The former elements are square-shaped through-holes with a side length of $L_{hole} = 5$ μm, required to provide shorter release etch paths for the removal of the sacrificial layer, as can be seen in Figure 5. The latter elements are polysilicon elements of $t_{dimple} = 0.75$ μm height, placed under the suspended diaphragm in order to limit the contact surface and reduce the plate stiction occurrence [13], as showcased in Figure 6. Moreover, a lateral opening of side $L_{open} = 50$ μm has been added to the top plate anchoring structure, in order to facilitate the bottom plate electrical routing to the sensor bonding PADS. As expected, the greater complexity of this later model provokes a significant increase of the simulation time required to characterize its bending versus force behavior in ANSYS. However, the "complex" FE model produces more accurate results, showcasing an increased sensor sensitivity to the applied load, caused by the presence of hole cavities on the top plate; as well as a realistic contact point between the plates obtained under the presence of dimple elements, which reduce the plates gap in $t_{dimple} = 0.75$ μm [13].

Table 2. Design specifications for the square sensor's analytical and finite element (FE) models.

Parameter	Analytic Model	FE Simple Model	FE Complex Model
Force (F)	0–3 µN	0–3 µN	0–3 µN
Top plate side ($a_{top} = b_{top}$)	410 µm	410 µm	410 µm
Bottom plate side ($a_{bottom} = b_{bottom}$)	410 µm	390 µm	390 µm
Plates gap (t_g)	2 µm	2 µm	2 µm
Top plates thickness (t_m)	2 µm	2 µm	2 µm
Number of holes	0	0	144
Number of dimples	0	0	169
Side aperture	0	0	50 µm

Table 3. Design specifications for the circular sensor's analytical and FE models.

Parameter	Analytic Model	FE Simple Model	FE Complex Model
Force (F)	0–3 µN	0–3 µN	0–3 µN
Top plate radius (a_{top})	220 µm	220 µm	220 µm
Bottom plate radius (a_{bottom})	220 µm	210 µm	210 µm
Plates gap (t_g)	2 µm	2 µm	2 µm
Top plates thickness (t_m)	2 µm	2 µm	2 µm
Number of holes	0	0	479
Number of dimples	0	0	267
Side aperture	0	0	50 µm

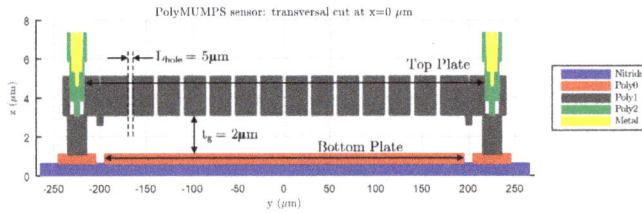

Figure 5. View of the transversal cut at $x = 0$ µm for a PolyMUMPS squared sensor, based on a top bottom plate configuration with sides $a_{bottom} = 390$ µm and $a_{top} = 410$ µm, respectively.

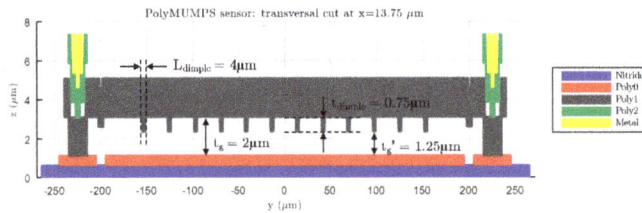

Figure 6. View of the transversal cut at $x = 13.75$ µm for a PolyMUMPS squared sensor, based on a top bottom plate configuration with sides $a_{bottom} = 390$ µm and $a_{top} = 410$ µm, respectively.

As a result of the aforementioned modelling, Figures 7 and 8 show the maximum bending estimation against the concentrated force applied to the sensors described in Tables 2 and 3, respectively. Both the analytical and FE responses are presented in each figure to ease the comparison between models. Additionally, the full scale error value is provided to quantify the mismatch between both models. According to the collected results, it can be stated that the analytical expressions (14) and (18) tend to underestimate the bending of the plates, when compared with the results of the FE models. However, because of their easy translation to the mathematical software (Matlab) and their fast evaluation, the analytical approach can still be considered an accurate-enough method to estimate the sensor behavior in an efficient way, previous to the FE model implementation in ANSYS. Alternative numerical approximation methods can be used to enhance the achieved exactitude, at the expense of immediacy, increased complexity, and higher computing requirements.

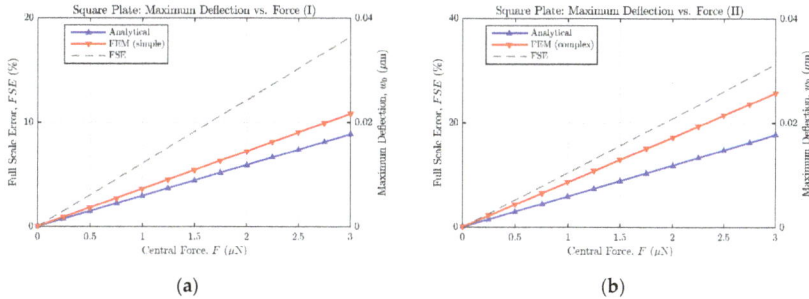

Figure 7. Maximum deflection versus central force for a fully clamped square plate. (**a**) Analytical model and finite element (FE) simple model results; (**b**) analytical model and FE complex model results.

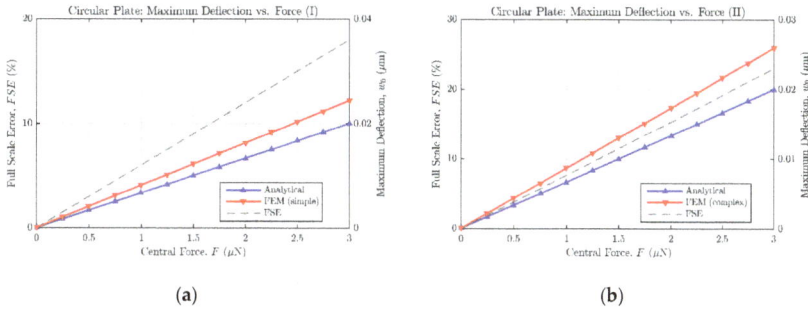

Figure 8. Maximum deflection versus central force for a fully clamped circular plate. (**a**) Analytical model and FE simple model results; (**b**) analytical model and FE complex model results.

2.2. CardioMEMS Design Tool

MEMS sensor design flow, as shown in Figure 9a, comprises three main stages. First, a rough sensor design proposal is obtained by the evaluation of a set of analytical expressions, while satisfying the initial design constraints. Second, the sensor realistic 3D structure is implemented in a modelling software for FE analysis; in order to perform the necessary simulations to achieve an accurate characterization of the device behavior. Finally, the third stage involves the sensor layout description in the corresponding technology layers being sent to the manufacturer. CardioMEMS Design (CMD) is a Matlab-based computer aided design (CAD) tool developed with the aim of automating the aforementioned design steps, as well as providing a friendly interface to guide the user through the sensor design process [14,15]. Thus, as exposed in Figure 9b, CMD provides as outputs the set of files required to export the designed sensor to ANSYS and Cadence Virtuoso working environments, to perform different FE analyses and build the sensor layer description, respectively.

The use of CMD to perform the complete design of a capacitive MEMS pressure sensor for ISR-monitoring iStents is described next, in order to present its functionalities and features. First, as can be seen in Figure 10, CMD requires the definition of various input parameters to provide an initial sensor design, including the selection of a fabrication technology included in the database, the preferred sensor topology, the maximum pressure to be borne by the transducer, and the desired width of the line used to physically connect the sensor to the bonding PADs.

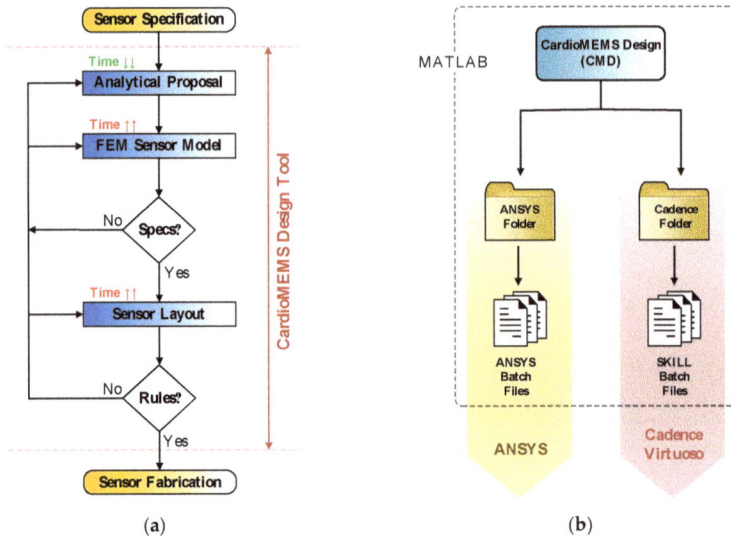

Figure 9. CardioMEMS Design (CMD) functionalities overview. (**a**) MEMS sensor design flow, highlighting those stages covered by CMD; (**b**) CMD output folders generated to export the designed sensor to ANSYS and Cadence Virtuoso.

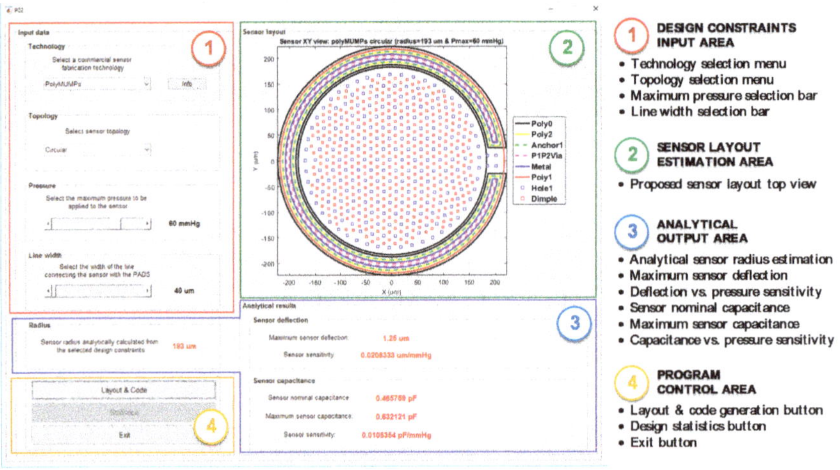

Figure 10. CardioMEMS Design (CMD) main screen, with its working areas and main functionalities displayed.

In the case of our prototype sensors, the selected MEMS technology has been PolyMUMPS by MEMSCAP, a mature and reliable surface-micromachining fabrication process, developed to accommodate a wide variety of MEMS structures. PolyMUMPS uses eight masks to define the topology of seven physical layers: three polysilicon layers, a metal (Au) layer, and two phospho-silicate glass (PSG) sacrificial layers [13]. It is important to keep in mind that the information about any new fabrication process must be added to the CMD database to allow for the use of that particular technology. Second, any MEMS pressure sensors intended to be used for mild ISR detection in a distal ramification of the pulmonary artery must face pressures in the range of $P = [0, 60]$ mmHg [16], while presenting the maximum achievable sensitivity to pressure changes. Thus, a circular sensor topology has been chosen initially, because of its higher sensitivity compared with a squared-shaped sensor of the same area [14]. Finally, a 40 μm wide Polysilicon line has been selected to physically connect the sensor to the bonding PADs. Table 4 includes a summary of the input parameters introduced in CMD's Design Constraints Area.

Table 4. CardioMEMS Design (CMD)-input parameters selected to perform the sensor design.

Parameter	Value
Fabrication technology [1]	PolyMUMPS
Diaphragm shape	Circular
Detectable pressure range	0 to 60 mmHg
Line to PAD width	40 μm

[1] The fabrication technology must be included in the program database.

Once the input parameters have been defined, CMD determines the optimum sensor radius a by evaluation of the analytical equations for the deflection of a circular or square plate under a uniformly applied pressure. Internally, the program considers that the peak deflection, the contact between the suspended top plate and the fixed back plate, is achieved when the maximum pressure is applied to the sensor, thus imposing an additional constraint of $w_0 = t_g$. As can be seen in Figure 10, the main performance parameters for the proposed sensor are displayed in the analytical output area of the program, including the sensor radius (a), its maximum deflection (w_0) and deflection versus pressure sensitivity ($S_P^{w_0}$), together with its nominal (C_0) and maximum capacitances (C_{MAX}) and the capacitance versus pressure sensitivity ($S_P^{C_s}$). Because of the relative simplicity of the analytical models used, every time one of the input parameters is modified, the output parameters are quickly recalculated and shown to the user. This allow the designer to evaluate different sensor technologies, topologies, and pressure ranges, getting an initial estimation of their performance in an efficient and agile way.

The *Layout & Code* bottom in the CMD main screen becomes active right after the initial design proposal has been presented to the user. By clicking on it, CMD internally performs a design adjustment; now taking into account the set of design rules provided by the technology manufacturer. Additionally, CMD considers the functional limitations from both ANSYS and Cadence Virtuoso while performing the design rearrangement, in order to guarantee its compatibility with both environments.

In the case of PolyMUMPS sensors, the optimum location for both holes and dimples is selected by CMD through the evaluation of a geometric distribution algorithm, based on Delaunay triangulation theory [17]. This algorithm minimizes the number of elements added to the structure, while guarantying the fulfillment of the design rules imposed by the technology. The rearranged sensor design is presented to the user in the sensor Layout Estimation Area, providing a top view of the layout displaying all the required technology layers for its fabrication, as shown in Figure 10.

Additionally, CMD saves relevant statistical information about the layers geometry. This information is accessible by the *Statistics* button in the program control area, as indicated in Figure 10, only after the *Layout & Code* operation has concluded.

Figure 11 includes a summary of the statistical data analyzed by the program for a PolyMUMPS-based sensor design. More concretely, Figure 11a provides the number of vertices used to polygonise the sensor defining layouts, which is a Cadence Virtuoso limiting operation factor. As can be noted, CMD fixes the number of vertices to a maximum value of 200 per layer, so Cadence Virtuoso constraints are not infringed. Moreover, CMD takes into account the spacing limitations between hole and dimple elements, defined in the provided PolyMUMPS documentation as $d_{hole} = [3, 30]$ μm and $d_{dimple} \leq 3$ μm, respectively [13]. The distance distribution for each type of element is displayed in Figure 11c,d; while Figure 11b indicates the total number of elements added to the basic sensor structure. This set of data is provided to the designer in order to quickly check that the design restrictions established by both the manufacturer and the software suites are thoroughly satisfied.

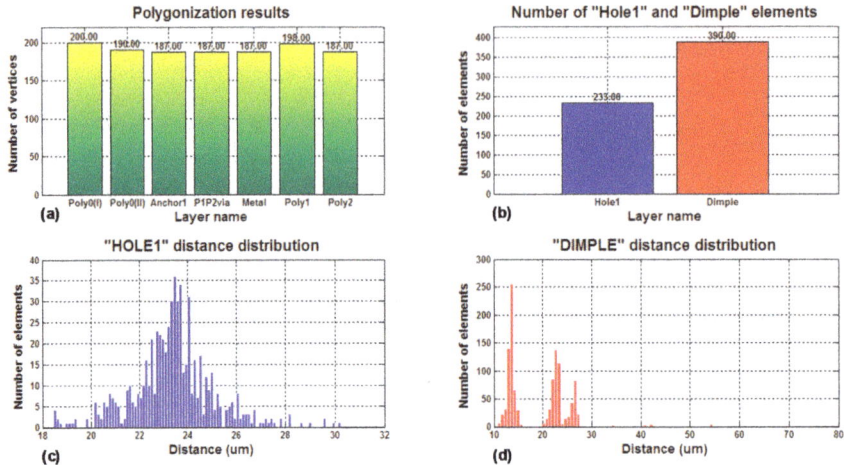

Figure 11. Sensor statistics provided by CMD, including (**a**) number of vertices for each fabrication technology layer; (**b**) number of "hole" and "dimple" elements included in the sensor structure; (**c**) distance distribution between "hole" elements; and (**d**) distance distribution between "dimple" elements.

As previously mentioned, and illustrated in Figure 9b, one of the main goals of CMD is to facilitate the sensor design translation to ANSYS and Cadence Virtuoso, so that the final stages of the design flow can be completed. Hence, after each design completion, CMD generates two output folders, denoted as *codeAnsys* and *codeCadence*, comprising a series of files suitable to automatically export the sensor design to those platforms.

The *codeAnsys* folder contains seven files, with three of them being auto-executable files, while the remaining four files have an auxiliary purpose. Those three main batch code files, once individually loaded in ANSYS, build the 3D solid model of the sensor, optimize its FE meshing, and perform different simulations to completely characterize the device behavior. For instance, one of the files configures the program solver to calculate the sensor deflection and equivalent capacitance for a uniformly applied pressure in the intended operation range of $P = [0, P_{MAX}]$ mmHg Similarly, a second main file sets the solver to evaluate the sensor deflection and capacitance under the presence of a central force load (F), similar to the one that can be applied by an AFM microscope operating in contact mode [6–9]; and described in detail in Section 2.1. Finally, the last main code file arranges a modal analysis to determine the natural frequencies of the structure.

In Figure 12, the central deflection and capacitance versus pressure simulation results are summed up. As can be observed, the simulation results given by ANSYS diverge slightly from those anticipated by the set of analytical expressions [14]. It must be acknowledged by the user that CMD proposes an initial rough sensor geometry based on the evaluation of analytical/numerical equations, because of their high computational efficiency. On the other hand, the CMD-rearranged design built in ANSYS presents higher complexity, mainly due to the addition of hole and dimple elements, responsible for reducing the stiffness of the movable plate [14] and limiting the effective gap distance to $t'_g = t_g - t_{dimple} = 1.25$ µm, respectively. Furthermore, as detailed in Figures 5 and 6, the realistic sensor model requires a moderately smaller backplate compared with the suspended one, which negatively affects the sensor nominal capacitance (Figure 12b). It can be perceived how the increased complexity of the structure contributes to locating the contact point in the range of measurable pressures; meaning that the initial sensor proposal underestimates the device sensitivity. However, for some applications, it can be desired to force the sensor operation entirely in contact mode, so a lineal capacitance response can be achieved [18]. The designer can take advantage of this behavior, being able to get exponential or linear capacitance versus pressure responses just by properly selecting the maximum detectable pressure at the beginning of the design flow.

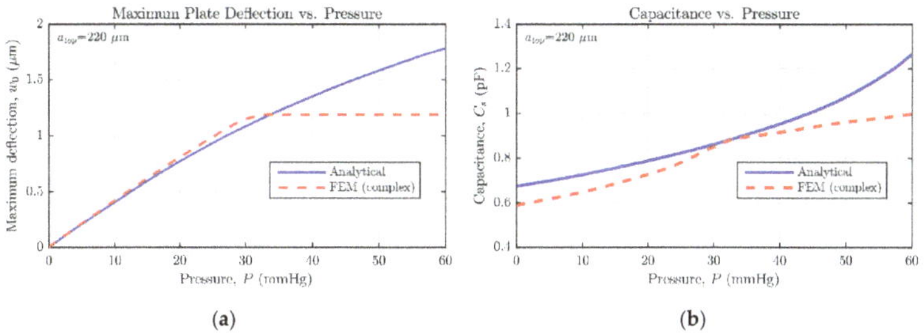

(a) (b)

Figure 12. Comparison between the analytical and FEM models for circular MEMS pressure sensor with top plate radius of $a_{top} = 220$ μm. (**a**) center deflection versus pressure response; (**b**) capacitance versus pressure response.

The *codeCadence* folder contains a unique auto executable SKILL batch file, supported by eighteen auxiliary files used to individually define the geometry of each fabrication layer. After being loaded in Cadence Virtuoso, the main batch file conducts the drawing of the necessary layers to create an adequate GDSII stream format file to be sent to the manufacturer. Figure 13a shows the layout view of the prototype sensor automatically built in Cadence Virtuoso's Layout Suite; where the bonding PADs are the only elements manually added by the designer. Besides, for comparison purposes, Figure 13b presents a scanning electron microscope (SEM) image of the fabricated sensor, defined by the layout in Figure 13a.

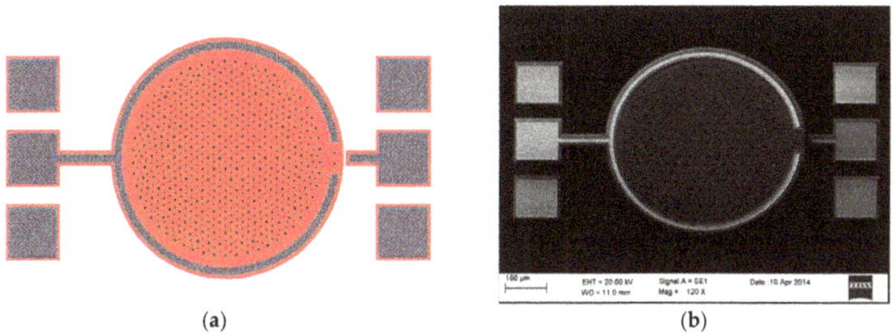

(a) (b)

Figure 13. Final design stages for a circular MEMS pressure sensor with top plate radius of $a_{top} = 220$ μm. (**a**) Cadence Virtuoso layout view, generated through the SKILL batch files provided by CMD; (**b**) scanning Electron Microscope (SEM) picture of the fabricated sensor.

3. Results

The deflection versus force response of two prototype PolyMUMPS square and circular pressure sensors, referred to as *S02_V03* and *S03_V03*, respectively, has been experimentally characterized using the atomic force microscope model XE-100 by Park Systems (Figure 14a). These measurements were developed at the Centro Tecnológico de Componentes, CTC (Technological Centre of Components) in Santander, where the AFM machine is located.

The microscope mounts a probe PPP-NCHR with a spring constant of nominal value $k_{AFM} = 42$ N/m, and a tolerance range of $[10, 130]$ N/m, as specified by the manufacturer.

The AFM has been configured to apply a maximum force of $F_{MAX} = 3$ μN to the sensor plate, in order to obtain a force (F) against piezo displacement (Z_{SCAN}) curve similar to the one displayed in Figure 14b. Both sensors have been measured a total of 20 times, and during each individual measurement, the piezo displacement is varied in 4096 steps; so the resulting data set can be considered statistically significant.

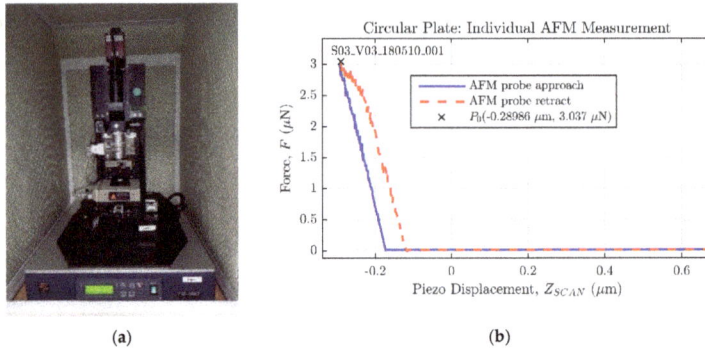

(a)   (b)

Figure 14. Experimental characterization equipment and sample measurement. (**a**) AFM model XE-100 by Park Systems; (**b**) sample contact mode measurement performed on a prototype circular MEMS sensor.

Figure 15a,b include the whole set of AFM measurements carried out for the square and circular sensors experimental characterization, respectively. Both graphs present the statistical data of sensor deflection for each applied force, displaying its average value, the 25th and 75th percentiles, and the extreme collected values. With this information, a least-squares fitting procedure has been applied to obtain a linear approximation for the aforementioned data, probing the maximum deflection to be proportional to the applied force as predicted by analytical and FE simulation.

Besides probing the linearity of the sensor response under a central force, the experimental results also demonstrate that maximum deflections for both geometries are almost equal, as predicted with the analytical and FE models. Although in a different scale, both graphs in Figure 8 show that each maximum deflection value obtained with the "complex" FE model for the circular sensor matches its square counterpart. Similarly, the abovementioned linear approximation of the measurements carried out with the AFM (Figure 15) demonstrates this same behavior by showing a similar proportionality constant between maximum deflection and force for both the square and the circular plate.

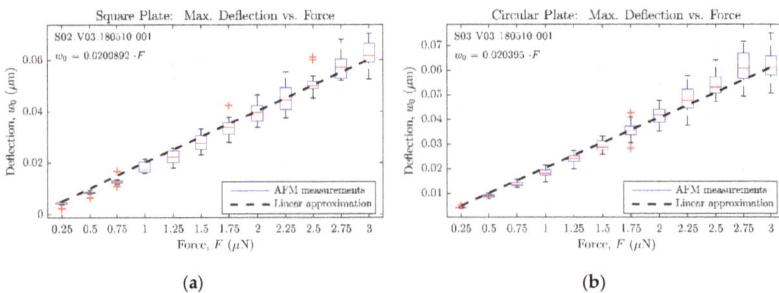

(a)   (b)

Figure 15. Experimental measurements of the sensor deflection versus a concentrated central force, assuming a AFM cantilever spring constant of $k_{AFM} = 42$ N/m. (**a**) For a square sensor; (**b**) for a circular sensor.

However, the experimental measurements bring out a higher sensor sensitivity than expected from both FE and analytical models. This result can be explained if the impact of the spring constant value is quantified, as done in Figure 16. These two graphs show how the selection of a spring constant value for the probe significantly affects the final result for the maximum deflection of both square and circular geometries of the sensor. As an example, the tolerance range set by the manufacturer goes from 10 to 130 N/m, but values below 25 N/m would lead to unreasonable negative deflection results.

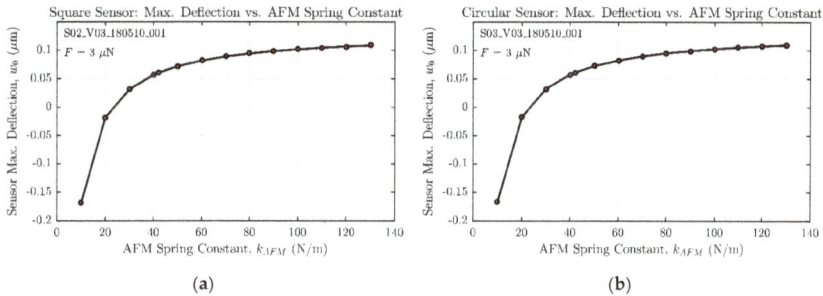

Figure 16. AFM cantilever spring constant influence over the estimation of sensor deflection. (**a**) Square sensor analysis; (**b**) circular sensor analysis.

Figure 17 provides an explanation for this equal behavior in spite of considering different geometries for the sensor. In this figure, the stress distribution under the same central force for both cases is depicted. As can be seen, this force is not big enough to induce significant stress outside the immediate vicinity of the plate center, where the geometrical constraints of both types of sensors have no effect and both behave as if they were circular.

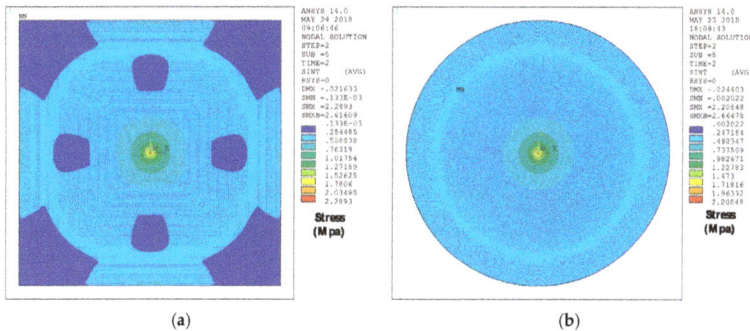

Figure 17. ANSYS results for the sensor stress distribution in MPa under an applied concentrated force of $F_{MAX} = 3$ µN at the top plate center. (**a**) Square sensor stress distribution; (**b**) circular sensor stress distribution.

4. Discussion

In this work a novel characterization method for prototypes of capacitive MEMS pressure sensors using an AFM has been presented. This approach takes advantage of a relatively common laboratory instrument, applied to a non-coated prototype for biomedical applications, where sensitivity and reliability are critical. Hence, this methodology helps to reduce the cost of the testing setup while ensuring accuracy and increasing reliability. The AFM has been configured in contact mode to apply a concentrated force on the center of the top plate of the sensor, in order to obtain a force versus piezo vertical displacement from which the maximum deflection of the plate under concentrated load can be determined.

Two prototypes of MEMS sensors with different geometries (circular and square) have been submitted to AFM characterization, and the resulting experimental data have been compared with simulation results developed on both analytical and FE models. Regarding these last ones, two different FE models with variable complexity have been developed to perform deflection versus force simulations and compare the resulting bending data to the response anticipated by the analytical expressions. A first FE model, referred as "simple", requires low computational time as it consists only of a flat square or circular plate fully clamped along its edges. The second FE model, named "complex", presents the exact same topology as the prototype sensor fabricated in PolyMUMPS technology and it includes hole and dimple elements as required by the manufacturer.

The complete design of both prototypes of capacitive MEMS pressure sensors for ISR follow-up has been developed with a Matlab-based computer aided design (CAD) tool called CardioMEMS Design (CMD). This software tool automates the design steps, as well as provides a friendly interface to guide the user through this process that comprises three main stages: rough sensor design based on analytical expressions satisfying the initial specifications; sensor realistic 3D FE model; and sensor layout description in the corresponding technology layers to be sent to the manufacturer.

In light of the experimental results obtained by the AFM, the predicted linear behavior of the maximum deflection as a function of a concentrated load has been demonstrated. Besides that, the data extracted from AFM measurements also confirms that deflection values for both geometries of the sensor are almost equal within the applied force range. The similar stress distribution in the vicinity of the plate center that was extracted by simulation on FE models explain that geometry-independent response.

Nevertheless, the experimental measurements show a deviation in the absolute values for the maximum deflection compared with what was expected from both FE and analytical models. In this regard, the calculation method for sensor displacement is highly dependent on the spring constant, as demonstrated in the Section 3. To overcome this limitation in obtaining accurate absolute deflection values, a reference stiffness calibration method for the AFM probe could be applied.

As possible lines for future work, besides the application of the aforementioned calibration method, it will be interesting to study how this methodology can be adapted to analyze the sensor behavior against fatigue-caused aging.

Author Contributions: Conceptualization, J.A.M., Y.L. and M.M.; Methodology, J.A.M., Y.L. and M.M.; Software, J.A.M.; Formal Analysis, J.A.M.; Investigation, J.A.M., Y.L. and M.M.; Writing—Original Draft Preparation, J.A.M. and Y.L.; Writing—Review & Editing, J.A.M., Y.L. and M.M.; Project Administration, M.M.; Funding Acquisition, J.A.M., Y.L. and M.M.

Funding: This research was funded by the Spanish Government's "Ministerio de Economía, Industria y Competitividad" under the joint projects TEC2013-46242-C3-2-P and TEC2013-46242-C3, co-financed with FEDER.

Acknowledgments: This work was carried out in collaboration with the Cardiology Department of the UHMV Hospital, Santander (Spain).

Conflicts of Interest: The authors declare no conflict of interest. The funders had no role in the design of the study; in the collection, analyses, or interpretation of data; in the writing of the manuscript, and in the decision to publish the results.

References

1. Organization for Economic Co-Operation and Development (OECD)/European Union (EU). *Health at a Glance: Europe 2016: State of Health in the EU Cycle*; OECD Publishing: Paris, France, 2016. [CrossRef]
2. Townsend, N.; Wilson, L.; Bhatnagar, P.; Wickramasinghe, K.; Rayner, M.; Nichols, M. Cardiovascular disease in Europe: Epidemiological update. *Eur. Heart J.* **2016**, *37*, 3232–3245. [CrossRef] [PubMed]
3. Takahata, K.; Gianchandani, Y.B.; Wise, K.D. Micromachined antenna stents and cuffs for monitoring intraluminal pressure and flow. *J. Microelectromech. Syst.* **2006**, *15*, 1289–1298. [CrossRef]

4. Chow, E.Y.; Chlebowski, A.L.; Chakraborty, S.; Chappell, W.J.; Irazoqui, P.P. Fully wireless implantable cardiovascular pressure monitor integrated with a medical stent. *IEEE Trans. Biomed. Eng.* **2010**, *57*, 1487–1496. [CrossRef] [PubMed]

5. Brancato, L.; Keulemans, G.; Verbelen, T.; Meyns, B.; Puers, R. An Implantable Intravascular Pressure Sensor for a Ventricular Assist Device. *Micromachines* **2016**, *7*, 135. [CrossRef]

6. Patil, S.K.; Çelik-Butler, Z.; Butler, D.P. Characterization of MEMS piezoresistive pressure sensors using AFM. *Ultramicroscopy* **2010**, *110*, 1154–1160. [CrossRef] [PubMed]

7. Alici, G.; Higgins, M.J. Stiffness characterisation of microcantilevers based on conducting polymers. In Proceedings of the SPIE—The International Society for Optical Engineering, Melbourne, Australia, 9–12 December 2008; Volume IV, pp. 726806-1–726806-9. [CrossRef]

8. Rollier, A.S.; Legrand, B.; Deresmes, D.; Lagouge, M.; Collard, D.; Buchaillot, L. Tensile stress determination in silicon nitride membrane by AFM characterization. In Proceedings of the 13th International Conference on Solid-State Sensors, Actuators and Microsystems, Digest of Technical Papers, TRANSDUCERS '05, Seoul, Korea, 5–9 June 2005; Volume 1, pp. 828–831. [CrossRef]

9. Pustan, M.; Dudescu, C.; Birleanu, C.; Zygmunt, R. Nanomechanical studies and materials characterization of metal/polymer bilayer MEMS cantilevers. *Int. J. Mater. Res.* **2013**, *104*, 408–414. [CrossRef]

10. Timoshenko, S.; Woinowsky-Krieger, S. *Theory of Plates and Shells*, 2nd ed.; McGraw-Hill: New York, NY, USA, 1959; ISBN 0070858206.

11. Ventsel, E.; Krauthammer, T. *Thin Plates and Shells: Theory, Analysis and Applications*, 1st ed.; Marcel Dekker, Inc.: New York, NY, USA, 2001; ISBN 0824705756.

12. Young, D. Clamped Rectangular Plates with a Central Concentrated Load. *J. R. Aeronaut. Soc.* **1940**, *44*, 350–354. [CrossRef]

13. Cowen, A.; Hardy, B.; Mahadevan, R.; Wilcenski, S. *PolyMUMPs Design Handbook*, 13th ed.; MEMSCAP Inc.: Durham, NC, USA, 2011.

14. Miguel, J.A.; Rivas-Marchena, D.; Lechuga, Y.; Allende, M.A.; Martinez, M. A novel computer-assisted design tool for implantable MEMS pressure sensors. *Microprocess. Microsyst.* **2016**, *46*, 75–83. [CrossRef]

15. Rivas-Marchena, D.; Olmo, A.; Miguel, J.A.; Martinez, M.; Huertas, G.; Yúfera, A. Real-Time Electrical Bioimpedance Characterization of Neointimal Tissue for Stent Applications. *Sensors* **2017**, *17*, 1737. [CrossRef] [PubMed]

16. Murphy, J.G.; Lloyd, M.A. *Mayo Clinic Cardiology Concise Textbook*, 2nd ed.; Wiley-IEEE Press: Chichester, UK, 2010; ISBN 0199915712.

17. Persson, P.-O.; Strang, G. A simple mesh generator in MATLAB. *Soc. Ind. Appl. Math.* **2004**, *46*, 329–345. [CrossRef]

18. Luo, X.; Gianchandani, Y.B. A 100 μm diameter capacitive pressure sensor with 50 MPa dynamic range. *J. Micromech. Microeng.* **2016**, *26*, 045009. [CrossRef]

micromachines

MDPI

Article

Encapsulation of NEM Memory Switches for Monolithic-Three-Dimensional (M3D) CMOS–NEM Hybrid Circuits

Hyun Chan Jo and Woo Young Choi *

Department of Electronics Engineering, Sogang University, Seoul 04107, Korea; jhc10337@naver.com
* Correspondence: wchoi@sogang.ac.kr; Tel.: +82-2-715-8467

Received: 30 May 2018; Accepted: 20 June 2018; Published: 23 June 2018

Abstract: Considering the isotropic release process of nanoelectromechanical systems (NEMSs), defining the active region of NEM memory switches is one of the most challenging process technologies for the implementation of monolithic-three-dimensional (M3D) CMOS–NEM hybrid circuits. In this paper, we propose a novel encapsulation method of NEM memory switches. It uses alumina (Al_2O_3) passivation layers which are fully compatible with the CMOS baseline process. The Al_2O_3 bottom passivation layer can protect intermetal dielectric (IMD) and metal interconnection layers from the vapor hydrogen fluoride (HF) etching process. Thus, the controllable formation of the cavity for the mechanical movement of NEM devices can be achieved without causing any damage to CMOS baseline circuits as well as metal interconnection lines. As a result, NEM memory switches can be located in any place and metal layer of an M3D CMOS–NEM hybrid chip, which makes circuit design easier and more volume efficient. The feasibility of our proposed method is verified based on experimental results.

Keywords: CMOS–NEMS; NEMS; NEM memory switch; encapsulation; M3D

1. Introduction

Complementary metal-oxide-semiconductor–nanoelectromechanical (CMOS–NEM) hybrid circuits have been researched intensively thanks to their unique advantages: low power consumption, high performance, low fabrication cost and high chip density [1–9]. Some pioneering experimental results of CMOS–NEM hybrid circuits have been reported [2,5]. They have NEM devices on the top of a chip or in CMOS back-end-of-line (BEOL) metal interconnection layers. For the implementation of monolithic-three-dimensional (M3D) CMOS–NEM hybrid circuits, the release process is important to form the atmospheric or vacuum environment for the mechanical motion of NEM memory switches whose operating mechanisms have already been explained elsewhere [1,2]. Generally, the release process is performed by using vapor hydrogen fluoride (HF) etching. By using the vapor HF etching, the inter-metal-dielectric (IMD) layers such as the tetraethyl orthosilicate (TEOS) layers, which surround NEM devices, can be effectively removed with high selectivity toward metal layers [10]. However, a conventional release process using vapor HF etching can cause catastrophic influences on IMD and metal interconnection layers because it is an isotropic etching process: NEM structures and adjacent metal interconnection lines collapse due to the widespread removal of IMD layers. Thus, as shown in Figure 1a,b, it is difficult to place the metal interconnection lines around NEM memory switches, which will be called the "dead zone" in this manuscript. The existence of the dead zone makes M3D CMOS–NEM hybrid circuit design difficult and volume inefficient.

To minimize the dead zone surrounding NEM devices, this manuscript proposes a novel CMOS-process-compatible encapsulation method as shown in Figure 1c. In the proposed method, NEM memory switches are encapsulated by alumina (Al_2O_3) bottom/top passivation layers. The TEOS lower/upper sacrificial layers encapsulated by the Al_2O_3 bottom/top passivation layers are selectively removed by vapor HF etching while the rest of the regions are protected. Thus, the controllable formation of a cavity is feasible for the mechanical movement of NEM devices without damaging CMOS baseline circuits and metal interconnect lines. From now, this cavity will be called the "active region" of a NEM memory switch. To sum up, because our proposed encapsulation method defines the active regions of NEM devices without generating dead zones, they can be placed in any metal interconnection layers. To confirm the proposed method, prototype encapsulated NEM memory switches are implemented.

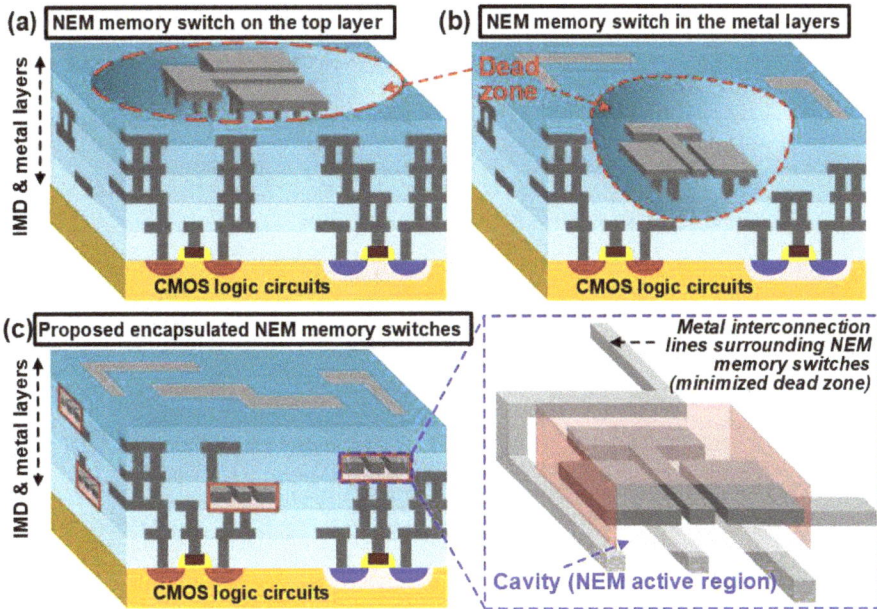

Figure 1. Conceptual views of (**a**) a nanoelectromechanical (NEM) memory switch only on the top layer, (**b**) a NEM memory switch in the CMOS back-end-of-line (BEOL) metal layers and (**c**) the proposed encapsulated NEM memory switches for monolithic-three-dimensional (M3D) CMOS–NEM hybrid circuits.

2. Encapsulation Process

Figure 2 shows the key process steps of the encapsulated nanoelectromechanical (NEM) memory switches. First, a 50-nm-thick silicon dioxide (SiO_2) layer is grown by wet oxidation. Then, a 500-nm-thick aluminum (Al) layer is sputtered and patterned by inductively coupled plasma (ICP) etching. The Al patterns correspond to the metal interconnect lines of CMOS baseline circuits. Third, a 500-nm-thick tetraethyl orthosilicate (TEOS) inter-metal-dielectric (IMD) layer is deposited and patterned by plasma-enhanced chemical vapor deposition (PECVD) and magnetically enhanced reactive ion etching (MERIE) processes, respectively, to define the active regions of NEM memory switches. Subsequently, a 200-nm-thick Al_2O_3 bottom passivation layer is deposited by a multisputtering process. The Al_2O_3 bottom passivation layer protects the metal interconnection lines and IMD layers from the following vapor hydrogen fluoride (HF) etch at atmospheric pressure [11–13]. Fifth, a 200-nm-thick TEOS layer is deposited as a lower sacrificial layer. Next, a 500-nm-thick Al layer is deposited and patterned to form NEM memory switches. During the patterning process, the 85-nm-wide airgap between the movable cantilever beam and selection lines is formed by a focus ion beam (FIB) process while the rest of the patterns are defined by a conventional stepper. Seventh, a 500-nm-thick TEOS layer is deposited and patterned as an upper sacrificial layer. It should be noted that the active regions of NEM memory switches are defined and filled by the lower and upper sacrificial layers. Eighth, a 200-nm-thick Al_2O_3 top passivation layer is deposited to encapsulate the active regions of NEM memory switches. Subsequently, small-sized etch holes are patterned on the Al_2O_3 top passivation layer by the FIB process. Tenth, the lower and upper TEOS sacrificial layers are removed through the etch holes by vapor HF etching at 40 °C and 15 min. Finally, a thick TEOS IMD layer is deposited on the Al_2O_3 top passivation layer to form the cavity surrounding NEM memory switches which acts as the active region. The encapsulated active regions are in the vacuum condition depending on TEOS deposition conditions. This encapsulation method is fully CMOS-process-compatible, which can be easily applied to the fabrication of M3D CMOS–NEM hybrid circuits.

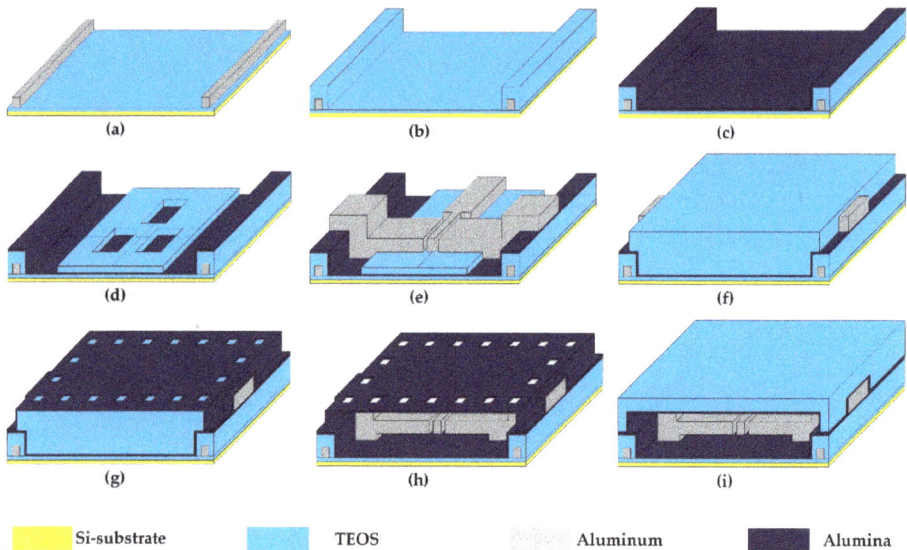

Figure 2. Key process steps of the encapsulated nanoelectromechanical (NEM) memory switches. (**a**) Al deposition and patterning for the formation of metal interconnection lines; (**b**) Tetraethyl orthosilicate (TEOS) deposition and patterning for inter-metal-dielectric (IMD) formation; (**c**) Al_2O_3 bottom passivation layer deposition; (**d**) Lower TEOS sacrificial layer deposition and patterning; (**e**) Al deposition and patterning for the formation of a NEM memory switch; (**f**) Upper TEOS sacrificial layer deposition and pattern; (**g**) Al_2O_3 top passivation layer deposition and etch hole formation; (**h**) Removal of the lower/upper sacrificial layers through etch holes by using vapor hydrogen fluoride (HF) etching; (**i**) TEOS deposition for cavity sealing.

For cavity formation, the etch holes should have the aspect ratio high enough to prevent TEOS from filling the cavity again through the etch holes. Figure 3 shows scanning electron microscopy (SEM, Thermo Fisher Scientific, Waltham, MA, USA) cross-sectional images of etch holes. In order to form the etch holes with various aspect ratios, two FIB process conditions have been adjusted: beam current and target diameter. The aspect ratio of the etch holes in Figure 3a–b are measured to be 0.79 (beam current = 50 pA and target diameter = 160 nm) and 1.01 (beam current = 10 pA and target diameter = 160 nm), respectively. It is interesting that two different layers are observed below the Al_2O_3 top passivation layer in those two cases. The former is a thin TEOS layer which is originated from the unwanted TEOS inflow through the etch holes. It is problematic in that it prevents the motion of a cantilever beam of a NEM memory switch. On the contrary, the latter results from the redeposition process during the FIB sample cutting process for SEM measurement, which does not exist in the main samples [14]. Thus, to suppress TEOS inflow, the aspect ratio of the etch holes needs to be increased. If the aspect ratio is increased up to 1.14 (beam current = 10 pA and target diameter = 80 nm) as shown in Figure 3c, no unwanted TEOS inflow is observed. Only the redeposition layer originated from the FIB sample cutting process is formed under the Al_2O_3 top passivation layer.

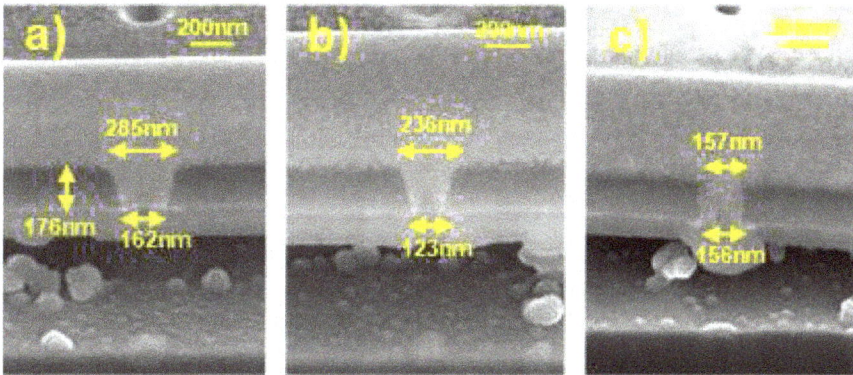

Figure 3. Cross-sectional scanning electron microscopy (SEM) images of etch holes with the variation of the beam current and target diameter of the focus ion beam (FIB) process. (**a**) Aspect ratio = 0.79 when beam current is 50 pA and target diameter is 160 nm; (**b**) Aspect ratio = 1.01 when beam current is 10 pA and target diameter is 160 nm. (**c**) Aspect ratio = 1.14 when beam current is 10 pA and target diameter is 80 nm.

3. Results and Discussion

Figure 4 shows the SEM images of the fabricated NEM memory switch encapsulated in a cavity. Figure 4a–f show the NEM memory switches before and after vapor hydrogen fluoride (HF) etching, respectively. The active region of the encapsulated NEM memory switch is formed well next to the metal interconnection lines, as shown in Figure 4. Figure 4b,c confirm that Al_2O_3 top and bottom passivation layers wrap the NEM memory switch and lower/upper tetraethyl orthosilicate (TEOS) sacrificial layers. Figure 4d–f show that the TEOS lower/upper sacrificial layers are successfully removed by vapor HF etching. In Figure 4e, it is confirmed that the sacrificial layers are completely removed by vapor HF without damaging the cavity regions. This forms the active region of the NEM memory switch, allowing activation between metal layers. Especially, Figure 4e shows the successful implementation of the NEM memory switch in a cavity. On the other hand, Figure 4f shows that the inter-metal-dielectric (IMD) layer out of the cavity is also removed by vapor HF etching, which means that the Al_2O_3 bottom passivation layer fails to protect the IMD layer from vapor HF etching. It is because vapor HF can penetrate into the Al_2O_3 layer following grain boundaries if the Al_2O_3 layer is formed by the sputtering process. Thus, in order to increase the film density of the Al_2O_3 passivation layer, an atomic layer deposition (ALD) process is used rather than a sputtering process. Figure 5a–d show the transmission electron microscopy (TEM) images of the test sample using a 20-nm-thick ALD-deposited Al_2O_3 layer before and after 1-, 5- and 15-min vapor HF etching at 40 °C, respectively. As predicted, it is observed that the SiO_2 IMD layer is completely protected by the ALD-deposited Al_2O_3 layer.

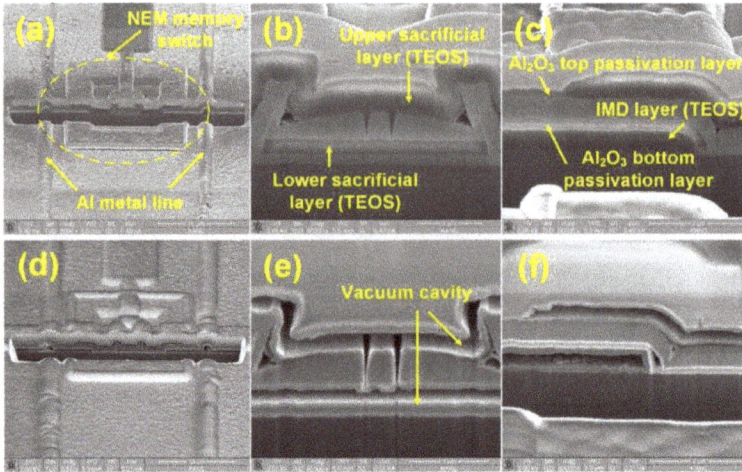

Figure 4. (**a**) Nanoelectromechanical (NEM) memory switch and metal interconnection lines; (**b**) NEM memory switch and (**c**) metal interconnection lines before vapor hydrogen fluoride (HF) etching; (**d**) NEM memory switch and metal interconnection lines; (**e**) NEM memory switch and (**f**) metal interconnection lines after vapor HF etching.

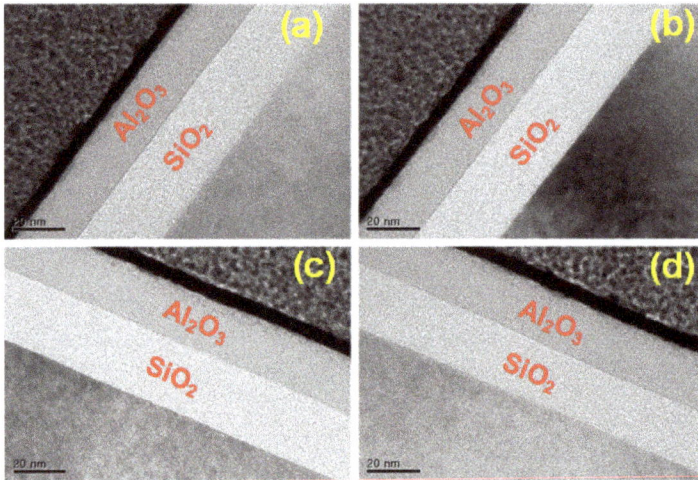

Figure 5. Transmission electron microscopy (TEM) images of an atomic layer deposition (ALD)-deposited Al_2O_3 layer (**a**) before and after (**b**) 1-min, (**c**) 5-min and (**d**) 15-min vapor hydrogen fluoride (HF) etching.

Figure 6 shows the current vs voltage curves of the fabricated NEM memory switch encapsulated in a cavity. It shows the reasonable nonvolatile switching operation between selection line 1 (L_1) and selection line 2 (L_2). The endurance cycle number is ~11 times due to the weak mechanical property of aluminum. In the first switching operation, the voltage difference between the movable cantilever beam and L_1 (V_{L1}) becomes higher than the pull-in voltage ($V_{pull-in}$), and then the movable cantilever beam is stuck onto L_1, which is called State 1. In this case, because the adhesion force (F_{ad}) is larger than the restoring spring force of the movable cantilever beam (F_r), the movable beam remains in contact with L_1 even when V_{L1} is 0 V [15]. Thus, the nonvolatile data signal storage can be achieved. In the second switching operation, the voltage difference between the movable cantilever beam and L_2 (V_{L2}) becomes higher than the switching voltage (V_{swit}), and then the location of the beam tip is changed from L_1 to L_2, which is called State 2. During the measurement, maximum current level was limited to suppress microwelding effects. Poor endurance cycle number can be improved by downscaling the dimension of NEM memory switches and changing beam materials [15,16].

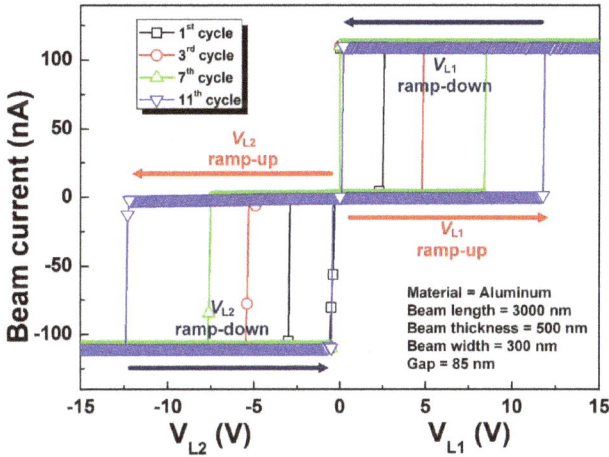

Figure 6. Current vs voltage curves of the fabricated nanoelectromechanical (NEM) memory switch encapsulated in a cavity.

4. Conclusions

In this work, a fabrication method to encapsulate an NEM memory switch for CMOS–NEM hybrid circuits is proposed by using a commercial CMOS process and materials. Specification of the stable encapsulated NEM memory switch is successfully confirmed based on the prototype fabrication and measurement results. By applying the proposed method confirmed in this work, the active regions of NEM memory switches can be formed without damaging CMOS baseline circuits as well as the metal interconnect lines. Because NEM memory switches can be located in any place and metal layer, the design of M3D CMOS–NEM hybrid chips can be easier and more volume efficient. It should be noted that our proposed encapsulation method can be applied to any kind of NEM device, including NEM switches, as long as they are fabricated by a CMOS backend process. For more uniform and reliable processes, a reduced-pressure vapor HF etcher can be used rather than the atmospheric-pressure vapor HF etcher used in this work. Therefore, the proposed fabrication process can lay the groundwork for commercialization of M3D CMOS–NEM hybrid circuits.

Author Contributions: W.Y.C. conceived and designed the experiments. H.C.J. performed the experiments. Both of the authors analyzed the data and wrote the paper.

Acknowledgments: This work was supported in part by the Sogang University Research Grant of 2017 (201710129.02), in part by the NRF of Korea funded by the MSIT under Grant NRF-2018R1A2A2A05019651 (Mid-Career Researcher Program), NRF-2015M3A7B7046617 (Fundamental Technology Program), NRF-2016M3A7B4909668 (Nano-Material Technology Development Program), in part by the IITP funded by the MSIT under Grant IITP-2018-0-01421 (Information Technology Research Center Program), and in part by the MOTIE/KSRC under Grant 10080575 (Future Semiconductor Device Technology Development Program).

Conflicts of Interest: The authors declare no conflict of interest.

References

1. Choi, W.Y.; Kim, Y.J. Three-Dimensional Integration of Complementary Metal-Oxide-Semiconductor-Nanoelectromechanical Hybrid Reconfigurable Circuits. *IEEE Electron Device Lett.* **2015**, *36*, 887–889. [CrossRef]
2. Kwon, H.S.; Kim, S.K.; Choi, W.Y. Monolithic Three-Dimensional 65-nm CMOS-Nanoelectromechanical Reconfigurable Logic for Sub- 1.2-V Operation. *IEEE Electron Device Lett.* **2017**, *38*, 1317–1320. [CrossRef]
3. Dong, C.; Chen, C.; Mitra, S.; Chen, D. Architecture and Performance Evaluation of 3D CMOS–NEM FPGA. In Proceedings of the System Level Interconnect Prediction Workshop (SLIP), San Diego, CA, USA, 5 June 2011; pp. 1–8.
4. Chong, S.; Lee, B.G.; Parizi, K.B.; Provine, J.; Mitra, S.; Howe, R.T.; Wong, P. Integration of nanoelectromechanical (NEM) relays with silicon CMOS with functional CMOS–NEM circuit. In Proceedings of the IEEE International Electron Devices Meeting (IEDM), Washington, DC, USA, 5–7 December 2011; pp. 701–704.
5. Muñoz-Gamarra, J.; Uranga, A.; Barniol, N. CMOS–NEMS Copper Switches Monolithically Integrated Using a 65nm CMOS Technology. *Micromachines* **2016**, *7*, 30. [CrossRef]
6. Muñoz-Gamarra, J.; Alcaine, P.; Marigó, E.; Giner, J.; Uranga, A.; Esteve, J.; Barniol, N. Integration of NEMS resonators in a 65nm CMOS Technology. *Microelectron. Eng.* **2013**, *110*, 246–249. [CrossRef]
7. Riverola, M.; Vidal-Alvarez, G.; Torres, F.; Barinol, N. 3-Terminal Tungsten CMOS–NEM Relay. In Proceedings of the Ph.D. Research in Microelectronics and Electronics (PRIME), Grenoble, France, 30 June–3 July 2014.
8. Harrison, K.L.; Clary, W.A.; Provine, J.; Howe, R.T. Back-end-of-line compatible Poly-SiGe lateral nanoelectromechanical relays with multi-level interconnect. *Microsyst. Technol.* **2017**, *23*, 2125–2130. [CrossRef]
9. Riverola, M.; Uranga, A.; Torres, F.; Barniol, N. Fabrication and characterization of a hammer-shaped CMOS/BEOL-embedded nanoelectromechanical (NEM) relay. *Microelectron. Eng.* **2018**, *192*, 44–51. [CrossRef]
10. Magis, T.; Ballerand, S.; Comte, B.; Pollet, O. Deep Silicon Etch for Biology MEMS Fabrication: Review of Process Parameters Influence versus Chip Design. In Proceedings of the SPIE MOEMS-MEMS, San Francisco, CA, USA, 9 March 2013; p. 826120A.
11. Witvrouw, A.; Bois, B.D.; Moor, P.D.; Verbist, A.; Hoof, C.V.; Bender, H.; Baert, C. Comparison between Wet HF Etching and Vapor HF Etching for Sacrificial Oxide removal. In Proceedings of the SPIE Micromachining and Microfabrication, Santa Clara, CA, USA, 25 August 2000; pp. 130–141.
12. Williams, K.R.; Gupta, K.; Wasilik, M. Etch Rate for Micromachining Processing-Part II. *J. Micromech. Syst.* **2003**, *12*, 761–778. [CrossRef]
13. Bakke, T.; Schmidt, J.; Friedrichs, M.; Völker, B. Etch Stop Materials for release by vapor HF etching. In Proceedings of the MicroMechanics Europe Workshop (MME), Göteborg, Sweden, 29 January 2005; pp. 103–106.
14. Winter, D.A.M.; Mulders, J.J.L. Redeposition Characteristics of Focus Ion Beam Milling for Nanofabricaiton. *J. Vac. Sci. Technol. B* **2007**, *25*, 2215–2218. [CrossRef]
15. Choi, W.Y.; Osabe, T.; Liu, T.J.K. Nano-electro-mechanical nonvolatile memory (NEMory) cell design and scaling. *IEEE Trans. Electron Devices* **2008**, *55*, 3482–3488. [CrossRef]
16. Soon, B.W.; Ng, E.J.; Qian, Y.; Singh, N.; Tsai, M.J.; Lee, C.K. A Bi-stable Nanoelectromechanical Nonvolatile memory based on van der Waals force. *Appl. Phys. Lett.* **2013**, *103*, 053122. [CrossRef]

micromachines

MDPI

Article

A Novel High-Precision Digital Tunneling Magnetic Resistance-Type Sensor for the Nanosatellites' Space Application

Xiangyu Li [1], Jianping Hu [1,*], Weiping Chen [2], Liang Yin [2] and Xiaowei Liu [2]

1   Faculty of Information Science and Technology, Ningbo University, Ningbo 315211, China; lixiangyu7410@sina.com
2   MEMS Center, Harbin Institute of Technology, Harbin 150001, China; weipingchen1@outlook.com (W.C.); 15B921019@hit.edu.cn (L.Y.); liuxiaowei3@outlook.com (X.L.)
*   Correspondence: hujianping2@nbu.edu.cn; Tel.: +86-0574-87600346

Received: 16 January 2018; Accepted: 6 March 2018; Published: 9 March 2018

Abstract: Micro-electromechanical system (MEMS) magnetic sensors are widely used in the nanosatellites field. We proposed a novel high-precision miniaturized three-axis digital tunneling magnetic resistance-type (TMR) sensor. The design of the three-axis digital magnetic sensor includes a low-noise sensitive element and high-performance interface circuit. The TMR sensor element can achieve a background noise of 150 pT/Hz$^{1/2}$ by the vertical modulation film at a modulation frequency of 5 kHz. The interface circuit is mainly composed of an analog front-end current feedback instrumentation amplifier (CFIA) with chopper structure and a fully differential 4th-order Sigma-Delta ($\Sigma\Delta$) analog to digital converter (ADC). The low-frequency $1/f$ noise of the TMR magnetic sensor are reduced by the input-stage and system-stage chopper. The dynamic element matching (DEM) is applied to average out the mismatch between the input and feedback transconductor so as to improve the gain accuracy and gain drift. The digital output is achieved by a switched-capacitor $\Sigma\Delta$ ADC. The interface circuit is implemented by a 0.35 μm CMOS technology. The performance test of the TMR magnetic sensor system shows that: at a 5 V operating voltage, the sensor can achieve a power consumption of 120 mW, a full scale of ±1 Guass, a bias error of 0.01% full scale (FS), a nonlinearity of x-axis 0.13% FS, y-axis 0.11% FS, z-axis 0.15% FS and a noise density of x-axis 250 pT/Hz$^{1/2}$ (at 1 Hz), y-axis 240 pT/Hz$^{1/2}$ (at 1 Hz), z-axis 250 pT/Hz$^{1/2}$ (at 1 Hz), respectively. This work has a less power consumption, a smaller size, and higher resolution than other miniaturized magnetometers by comparison.

Keywords: MEMS; interface circuit; chopper instrumentation amplifier; Sigma-Delta

1. Introduction

The earth is a huge magnetic source and scatters around the weak magnetic field (about 50 μT). So, there is a specific relationship between the size, direction of the magnetic field, and geographical position. We can achieve the high-precision GPS navigation by obtaining accurate geomagnetic field information. It is strategic and tactical for concealed combat equipment (such as submarines, stealth aircraft, etc.) [1]. The ferromagnetic objects (such as ore, magnetic conducting metal, armored vehicles, warships, submarines, etc.) can change the distribution of the geomagnetic field and generate anomaly magnetic field. If we can accurately measure the anomaly magnetic field, we can get the location, size, and other information of the target object [2]. As shown in Figure 1, the magnetic detection system plays an important role in the nano-satellite, unmanned aerial vehicle antisubmarine, ammunition fuze, geological exploration, mine clearance, and traffic monitoring. In the geomagnetic field detection, the signal amplitude of geomagnetic field and anomaly magnetic field are very weak, the signal frequency is very low (about 1 Hz).

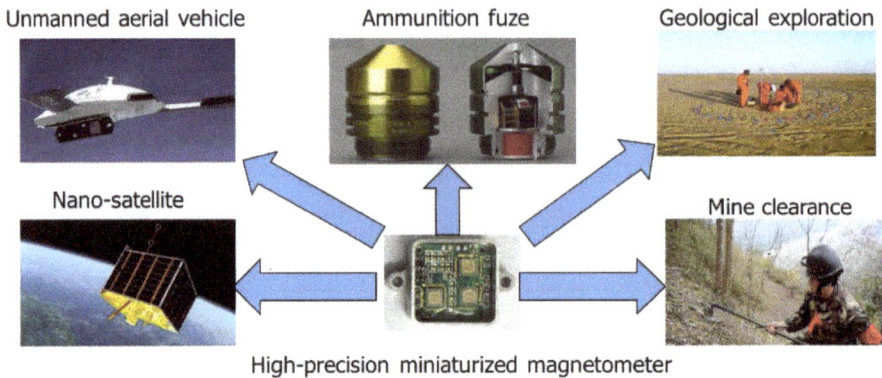

Figure 1. The high-precision miniaturized magnetometer in military and civilian field.

Miniaturization high-performance magnetometers based on AMR (anisotropic magnetic resistance), GMR (giant magnetic resistance), and TMR (tunneling magnetic resistance) are widely used in military and civilian field. It is difficult to improve the change rate of magneto-resistance based on the principle of anisotropic scattering. TMR based on tunneling current has a lager change rate of magneto-resistance compared with GMR and AMR. The TMR magnetometer has a higher sensitivity and a wider linear range as the third generation magneto-resistance sensor, but it has a lager noise at low-frequency than GMR and AMR because of tunneling effect. At present, the research on TMR sensors still stays on the study of sensitive surface materials. The research on reducing $1/f$ noise of TMR sensor and digital interface Application Specific Integrated Circuit (ASIC) for TMR sensors has not been reported. The study of a digital magnetometer based on TMR effect is significant. In the nanosatellites' space application field, magnetic sensors are widely used for attitude control. Because of miniaturization and low power consumption the magnetic resistance-type sensor is selected in most nanosatellites [3]. Magnetic sensors based on very large scale integration (VLSI) can combine with MEMS magnetic sensitive element. Honeywell manufactures several sensors based on AMR: one-axis HMC1021, two-axis sensors (HMC1022), and three-axis analog output (HMC1043) and digital output (HMR2300) with an integrated Application Specific Integrated Circuit (ASIC) [4–6]. For texample, in 2004, the ION-F (Ionospheric Observation Nanosatellie Formation) mission consists of three nanosatellites that were built at Utah State University and University of Washington. The four one-axis HMC1021 sensors are used for the attitude determination in NANOSAT-01 and NANOSAT-1B. This sensor has a linearity error of 0.4% FS, a noise density of 48 nV/Hz$^{1/2}$ and a sensitivity of 1 mV/V/Guass [7]. In 2009, the three-axis HMC1043 was applied in Spanish OPTOS satellite. It has a resolution of 13 nT and a noise density of 50 nV/Hz$^{1/2}$. The power consumption is increased to a total number of 500 mW.

As shown in Figure 2, the nano-satellites applied the Honeywell HMR2300 sensor: DawgStar satellite (University of Washington, Seattle, WA, USA), USUSat (Ohio State University, Columbus, OH, USA) and HokieSat (Virginia Tech, Blacksburg, VA, USA) [8]. The three-axis magnetometer HMR2300 with digital output can directly communicate with the computer. Three-axis independent structure can directionally detect the magnetic field of x-axis, y-axis, z-axis. The change voltage signal is amplified by the front-stage amplifier and converted to the digital signal by a 16-bit ADC. The sampling rate of input data, output format, average reading, and zero-bias offset can be set. The measure range is up to ±2 Gauss; the resolution is 6.7 nT; the sampling speed is 10–157 sampling points/s the three-axis digital output is BCD or binary code; the baud rate can be chosen 9600 or 19,200; the serial port can be chosen standard RS485 or RS232 for single-point reading; the volume is $70 \times 37 \times 24$ mm^3; and, the power consumption is 400 mW. It has a higher resolution and a better linearity error of 0.1–0.5% FS, but it has a big volume because of detection circuit and digital processing circuit [9,10]. Because of low sensitivity and linearity based on AMR and low-performance detection circuit in HMR2300, the precision of the magnetometer cannot be improved. But, in TMR sensors, the $1/f$ noise problem should be solved. The $1/f$ noise that is caused by a magnetic mechanism is always an important factor in affecting low-frequency magnetic detection capability of TMR sensors. According to the recently domestic and foreign research reports, high-frequency modulation method based on magnetic signal is the majority method to reduce $1/f$ noise in the miniaturized magnetoresistive sensor. In most of the micro/nano-satellite satellites miniaturization magnetic sensors are equipped with discrete devices, which have large volume and high power consumption. According to previous studies, the researchers used MgO materials as the barrier layer to improve the sensitivity of the TMR sensor, but the low-frequency noise cannot be reduced. The periodic modulation of the magnetic elements in the sensor at the chopping frequency have been proposed and tested for AMR and GMR. Here, we investigate the application of chopping techniques for TMR magnetic elements. The research on high performance digital interface ASIC for TMR sensors has not been reported. The study of high-precision miniaturized three-axis digital magnetic sensor system is necessary and significant.

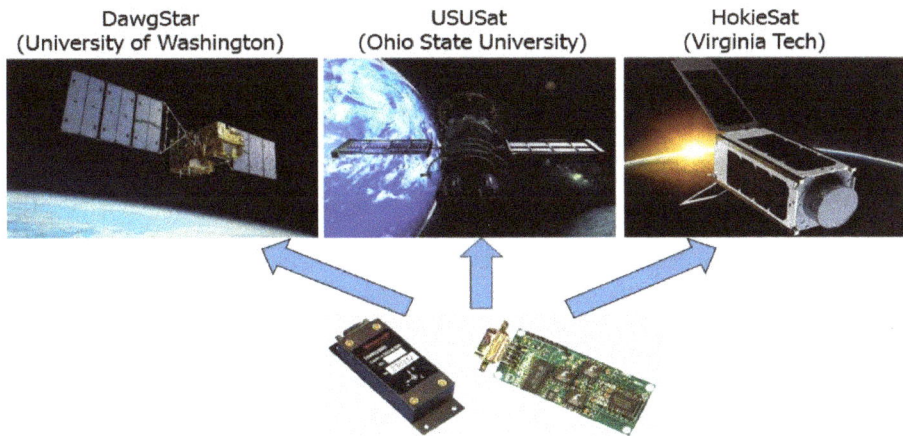

Figure 2. The HMR2300 magnetometer in Nano-satellite field.

In this paper, we propose a high-precision miniaturized three-axis digital TMR magnetic sensor for nanosatellites' space application. The TMR sensor element can achieve a background noise of $150\,\mathrm{pT/Hz^{1/2}}$ by the vertical modulation film. The analog front-end interface and digital ASIC chip for the tunneling magneto-resistive sensor are implemented by 0.35 μm mixed signal 5 V CMOS technology. The performance test of the TMR magnetic sensor system shows that: lower power consumption (120 mW), higher resolution ($250\,\mathrm{pT/Hz^{1/2}}$ (at 1 Hz)), better linearity (lower than 0.2% FS), and smaller volume ($25 \times 25 \times 10\,\mathrm{mm^3}$) than other miniaturized magneto-resistance sensor.

The tunneling magneto-resistance sensor element, current feedback instrumentation amplifier with chopper technique, Sigma-Delta modulator and digital decimation filter are introduced and designed in Section 2. In Section 3, we show a vertical modulation film structure and a detailed ASIC interface circuit. The performance can be improved by chopper technique, offset reduction loop technique, and dynamic element matching. The performance parameters of ASIC are tested by the experiments. Finally, Section 4 concludes the study of MEMS/TMR three-axis integrated magnetometer prototype and testing results, which shows that the performance of miniaturized TMR digital magnetometer has great advantages in the application of nano-satellite field.

2. Materials and Methods

2.1. Materials

The tunneling magneto-resistance sensor element with the multilaminar structure is from Multidimension Technologies (Suzhou, China). The TMR sensor interface circuit is fabricated by 0.35 μm CMOS process and cooperated with Shanghai Huahong Integrated Circuit (Shanghai, China).

2.2. Tunneling Magnetic Resistance-Type Sensor Element

The sensitive structure part of tunneling magneto-resistive sensor mainly consists of Pinning Layer, Tunnel Barrier, and Free Layer. The pinning layer is composed of a ferromagnetic layer and an anti-ferromagnetic layer (AFM Layer). The exchange coupling between the ferromagnetic layer and the anti-ferromagnetic layer determines the direction of the magnetic moment of the ferromagnetic layer; the tunneling barrier layer is usually composed of MgO or Al_2O_3, located in the upper part of the anti-ferromagnetic layer [11]. As shown in Figure 3a, the arrows represent the direction of the magnetic moment of the pinning layer and the free layer. The magnetic moment of the pinning layer is relatively fixed under the action of the magnetic field. The magnetic moment of the free layer is relatively free and rotatable to the magnetic moment of the pinning layer, and it will turn over with the change of the magnetic field. The typical thickness of each film layer is between 0.1 and 10 nm [12]. The magnetic sensor system is mainly composed of magnetic resistance-type sensitive element and CMOS readout integrated circuit. The sensitive element concludes 32 magnetic tunneling junctions (MTJ). The unit area resistance value RA of the magnetic tunneling junction is 2.5 kΩ/μm^2, the area of magnetic tunneling junctions is 50 μm^2. In this paper, the thickness of free layer/barrier layer/pinning layer is 10/1/10 nm. The multilayer structure of MTJ is Ta/Ru/Ta/PtMn/CoFe/Ru/CoFeB/MgO/CoFeB/NiFe/Ru/Ta, which structure is cooperated with Multidimension Technology Company. The thin film is deposited by magnetron sputtering. The MgO material is used as the barrier layer so that TMR element is more sensitive and higher resolution [13–16]. The Wheatstone bridge configuration is composed of four active TMR arrays that are applied by the thin film process, as shown in Figure 3a. The three-axis TMR sensitive element is built by stereoscopic orthogonal package.

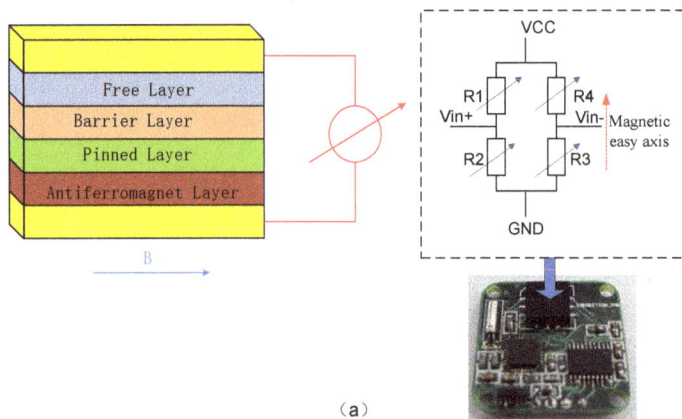

(a)

Figure 3. *Cont.*

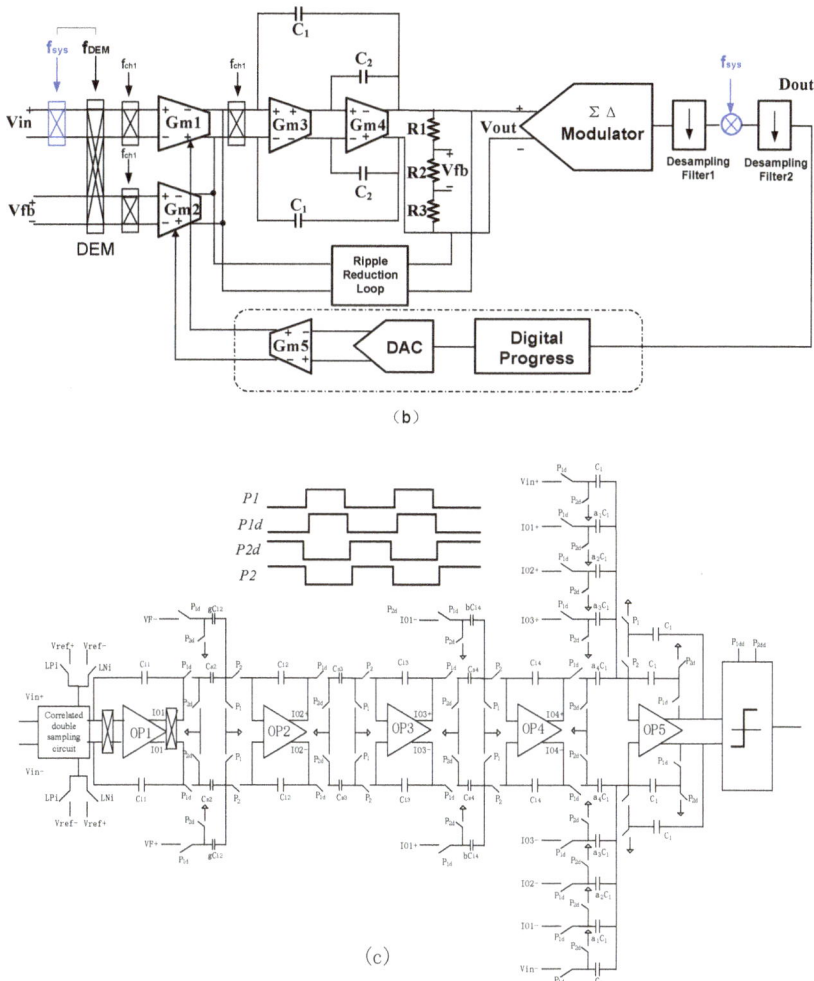

Figure 3. (**a**) Tunneling magnetic resistance-type sensitive structure, the physical printed circuit board (PCB) diagram is underneath the blue arrow (**b**) Three-axis digital tunneling magnetic resistance-type (TMR) sensor interface circuit System block diagram, the closed-loop feedback control of digital progress is shown in the dotted box (**c**) The fully differential 4th-order Sigma-Delta modulator.

2.3. Instrumentation Amplifier with Chopper Technique

The output signal of TMR sensors is at a low-frequency (about 1 Hz) and a millivolt range. Therefore, they need amplifiers to boost such a signal to be compatible with the input ranges of Sigma-Delta Analog-to-Digital Converters. Although the differential output voltage signal of the TMR sensors (V_{in}) can be as small as a few millivolts, the common-mode (CM) voltage V_{CM} depending on the application can be much larger and even vary at the range of a few volts during the operation, as shown in Figure 3b. To accommodate this variable CM voltage, an Instrumentation Amplifier is generally used for the read-out circuit of the sensors [17–19]. To accurately process the millivolt-level signal of the TMR sensor, the input referred error of the current feedback instrumentation amplifier (CFIA) should be at the microvolt or nanovolt-level [20–24]. To reduce $1/f$ noise, the chopper technique is applied in the circuit. To cope with varied CM voltage, the CFIA should have a common-mode rejection ratio (CMRR) greater than 120 dB. Furthermore, this CFIA is critical since it determines the overall performance of the read-out IC. To sum up, the main functions of this amplifier:

- Amplify the weak differential voltage (V_{in});
- Low offset, low noise and low corner frequency (<5 mHz);
- High common-mode rejection ratio (>120 dB); and,
- High input impedance for TMR sensor.

2.4. Sigma-Delta Modulator and Digital Decimation Filter

Figure 3b shows the ASIC part of the TMR magnetic sensor system. The output of the CFIA is digitized by a fully differential 4th-order Sigma-Delta ADC, which consists of a $\Sigma\Delta$ modulator and a decimation filter [25–27]. The circuit structure and sequence diagram of the fully differential modulator is as shown in Figure 3c. In order to achieve low noise and low offset, the chopper and correlated double sampling technique are both applied in the first stage integrator. The modulator can achieve a better noise suppression performance at low frequency [28,29]. The one-bit quantizer is achieved by the dynamic comparator. The output of the comparator is as a control signal to control feedback reference voltage V_{ref+} and V_{ref-} in the first stage integrator. As shown in Figure 3c, wherein P1 and P2 are the two-phase non-overlapping clock, P1 is active-high, P2 is active-low. The shutdown time of P1d is later than P1, The shutdown time of P2d is later than P2, it can effectively suppress the influence of charge injection and clock-feedthrough in the switched-capacitor circuit. The fully differential structure can effectively suppress even harmonics of $\Sigma\Delta$ ADC [30].

3. Result and Discussion

3.1. Noise Matching

Noise matching between the TMR sensor element, the instrumentation amplifier and the $\Sigma\Delta$ modulator is important effect on the precision of the TMR sensor system. The CFIA is critical in the interface circuit since it mainly determines the overall noise performance. The equivalent input noise of CFIA should be less than or equal to the TMR sensor element. To maintain the signal to noise ratio (SNR) of CFIA, the target resolution of the Sigma-Delta ADC is more than 18 bits and the background noise is less than −140 dB. The power spectrum density of each part is shown in Figure 4. The closed-loop gain of CFIA is 26 dB.

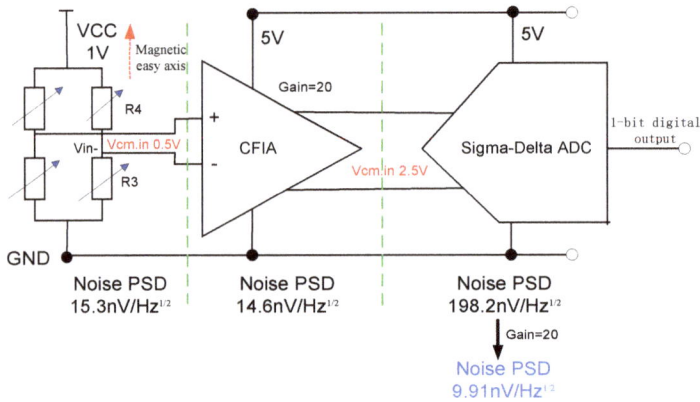

Figure 4. The noise performance of TMR sensor element, instrumentation amplifier and Sigma-Delta modulator.

3.2. Noise Characteristics of CFIA with TMR Sensor Element

The $1/f$ noise caused by magnetic mechanism is always an important factor in affecting low-frequency magnetic detection capability of TMR sensor. Here, we used the high frequency modulation method based on magnetic signal to reduce the low frequency $1/f$ noise of TMR sensor element. The flux modulation structure is as shown in Figure 5. The TMR sensor element is achieved to modulate by the vertical modulation film. The vertical modulation film based on the dynamic magnetic signal idea can directly modulate the measured magnetic field by the high-frequency vibratory machine-electric-magnetic microstructures. When the vertical modulation film is close to the TMR sensitivity element, the magnetic line of force is easier to pass through the vertical modulation film. The magnetic field will rapidly weaken near the TMR sensitivity element, as shown in Figure 5b. On the other hand, the vertical modulation film is far away from the TMR sensitivity element. The magnetic line of force tends to pass through the TMR sensitivity element. The shunting action of the vertical modulation film will decrease and the magnetic field near the TMR sensitivity element will be restored. Therefore, when the driving structure is in the case of high-frequency vibration, the periodic high-frequency vibration of the vertical modulation film is accompanied by the driving structure. The magnetic field near the TMR sensitivity element will be modulated. The TMR sensitivity element can detect an alternating magnetic field at this time because of the periodic high-frequency vibration. The finite element simulation of magnetic field with the vertical modulation film is shown in Figure 5b,c. The TMR sensor element can achieve a background noise of 150 pT/Hz$^{1/2}$ at a modulation frequency of 5 kHz.

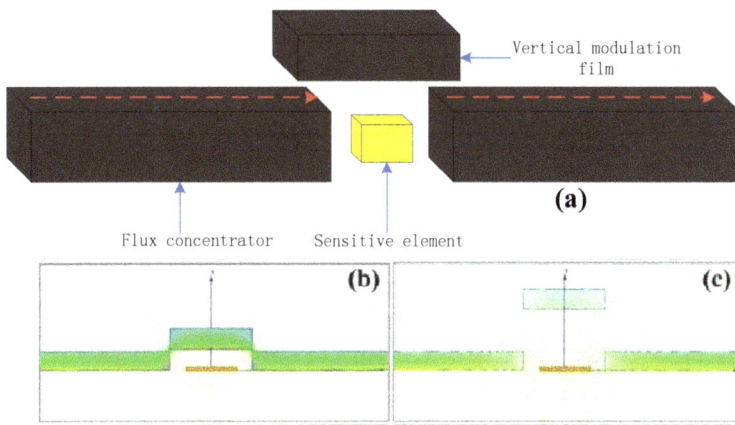

Figure 5. (**a**) The flux modulation micro-structure (**b**) TMR sensor with flux concentrators and choppers at OFF position (**c**) The choppers at ON position, the green line represents the size and direction of the magnetic field in the simulation.

3.3. Noise Characteristics of CFIA with Chopper Technique

Because the low-frequency $1/f$ noise of TMR sensors is the main noise, we use the vertical modulation film structure to reduce $1/f$ noise of the TMR element. But, in the circuit the $1/f$ noise of CFIA is large at low-frequency. This chopper method is effective. The relationship between the cutoff frequency of the chopper constituted of the analog switch and the input capacitance:

$$\frac{1}{2\pi R_S C_G} = \frac{1}{2\pi \mu_n C_{ox} C_G \frac{W}{L}(V_{GS} - V_{TH})} = 8f_{chop} \tag{1}$$

After the voltage noise of the analog switch is modulated by chopper-stabilized. The high frequency (f_{chop}) noise is modulated to low frequency, the equivalent input voltage noise density can be shown:

$$\overline{V}_n(f) \approx \mu_n C_{ox} \frac{W}{16\pi^2 L C_G (f - f_{chop})} \sqrt{\frac{K_f}{WLC_{OX}(f - f_{chop})}} \cdot V_{OS} \tag{2}$$

$$\frac{W}{L} = \frac{1}{16\pi\mu_n C_{ox} C_G (V_{VCC} - V_{in+} - V_{TH}) f_{chop}} \tag{3}$$

Equations (2) and (3) shows that the relationship between the voltage noise density of analog switches, the chopper stabilized amplifier's input offset voltage and the size of analog switches. At the same time, the minimum area of the analog switch is limited. In terms of tunneling magneto-resistive sensors' weak signal, the $1/f$ noise and KT/C noise are considered in the design of TMR interface ASIC [31–33]. In order to suppress the excessive noise, high-voltage CMOS technology of the high-amplitude clock feed-through and charge injection circuit are applied. In addition, the factors affecting the noise characteristics of the chopper switches are as below: charge leakage, parasitic capacitance, IC substrate coupling noise, voltage stability of the drive signal, and the external electric field sensitive electrodes [34,35]. The factors have been considered and optimization. In order to further reduce offset, system-level chopper can also suppress the CFIA's $1/f$ noise. The $1/f$ noise corner frequency is reduced from 10 to 0.3 Hz by the input-stage chopper. As long as the system-level chopper frequency is more than 10 kHz, the residual $1/f$ noise is suppressed. The simulating noise spectrum of various $1/f$ noise suppression techniques is shown as Figure 6b. The combination of the input-stage chopper and system-level chopper can achieve the best $1/f$ noise characteristics in the interface circuit. This front-end readout circuit can achieve mHz-level corner frequency and nanovolt-level offset. To improve gain accuracy, the dynamic element matching (DEM) is used for averaging out the mismatch between the input and feedback transconductors. The input-stage chopper frequency is 30 kHz, the resulting ripple is suppressed by a continuous-time ripple reduction loop (RRL), as shown in Figure 6a. The spectrum analyzer HP35670A (Hewlett-Packard, Palo Artaud, CA, USA) shows that the closed-loop gain is 26 dB, the unit gain bandwidth is 50 kHz (the signal bandwidth is 0–1 kHz), and the equivalent input noise density is 14.6 nV/Hz$^{1/2}$, as shown in Figure 6c.

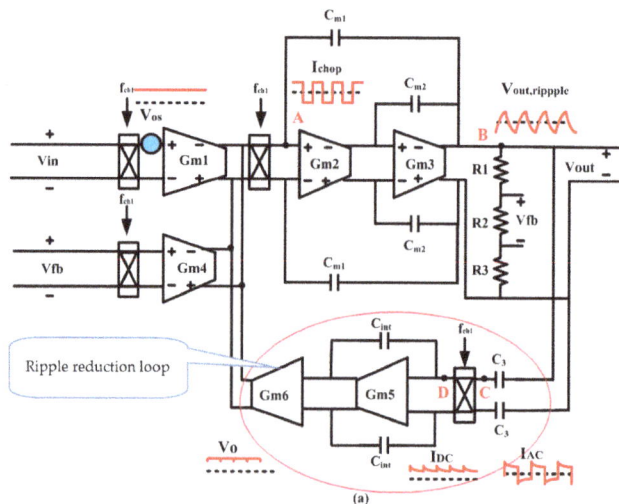

Figure 6. *Cont.*

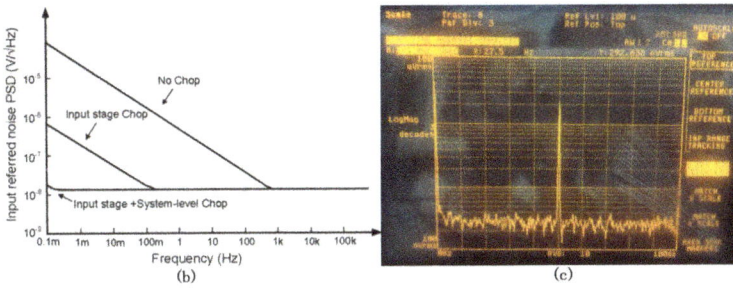

Figure 6. (**a**) Current feedback instrumentation amplifier (CFIA) circuit structure with ripple reduction loop (**b**) Simulating noise spectrum of with various noise chopper techniques (**c**) The power spectrum noise of CFIA.

The corner frequency of operational amplifier $1/f$ noise is usually about 10 kHz. To achieve a corner frequency of 1 mHz, the chopper technology of input-stage amplifier is applied. Meanwhile, the corner frequency of the second-stage equivalent to the first-stage should also be less than 1 mHz. Therefore, the input-stage DC gain A_{01} is required:

$$A_{01} \geq 20 \log_{10} \frac{10 \text{ kHz}}{1 \text{ mHz}} = 140 \text{ dB} \tag{4}$$

At the same time, the larger DC gain can effectively suppress the noise and the nonlinearity of the post-stage. The design of input-stage transconductance Gm1 and the current feedback transconductance Gm2 is particularly important, which determines the overall performance of the CFIA. To get higher gain accuracy, the circuit and layout are designed to satisfy the size and symmetry. The circuit structure of the input-stage chopper amplifier is shown in Figure 7a. A fully differential folded cascode structure with common-mode feedback is applied in the input-stage circuit. The circuit uses gain-booster (V_n and V_p) technology to improve the gain, and the circuit structure of V_n and V_p is shown in Figure 7b,c.

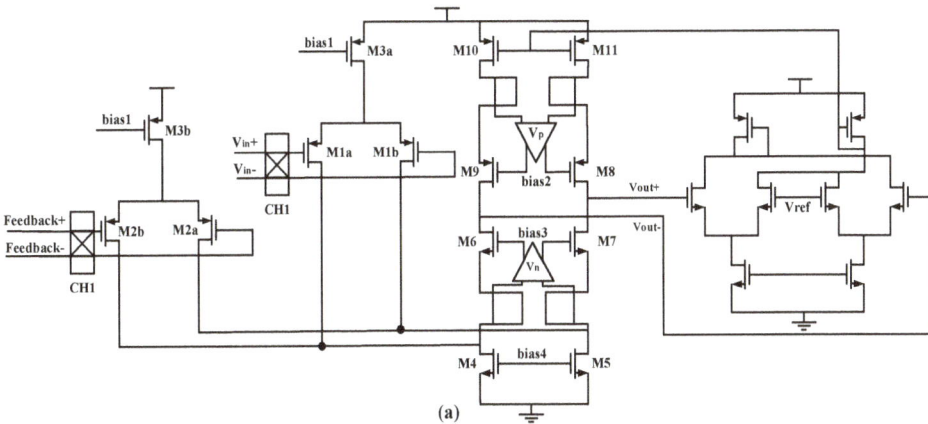

Figure 7. *Cont.*

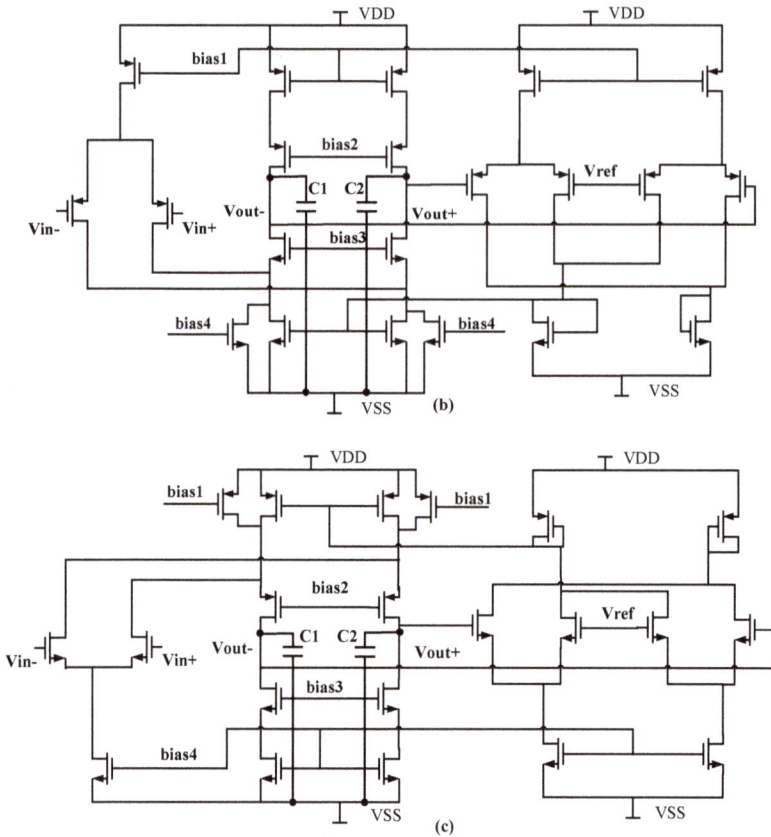

Figure 7. (**a**) Input stage transconductance operational amplifier with common-mode feedback circuit (**b**) Gain booster V_n; and, (**c**) Gain booster V_p.

3.4. Noise Characteristics of Sigma-Delta Modulator

In order to maintain the SNR of CFIA, the noise performance of the Sigma-Delta modulator at an over sampling ratio (OSR) of 128 is as shown in Figure 4. The target performance of Sigma-Delta modulator: a background noise of less than −140 dB within a 1 kHz signal bandwidth; a SNR of more than 110 dB; an effective number of bits more than 18 bits; a harmonic distortion of less than −110 dB. To obtain a conversion time less than 0.2 s, the required sampling frequency is only 30 kHz, which is equal to the chopper frequency of the CFIA. In order to test the noise characteristics of Sigma-Delta modulator, the Sigma-Delta modulator is implemented by standard 0.35 μm CMOS process. The testing system of the Sigma-Delta modulator is as shown in Figure 8a in this paper. The 5 V power supply is provided by Agilent 3631A (Agilent Technologies Inc, Santa Clara, CA, USA); a clock control signal is provided by the Tektronix AFG3102 function signal generator (Tek Technology Company, Shanghai, China); the analog input signal is provided by the Agilent 35670 (Agilent Technologies Inc, Santa Clara, CA, USA); the digital output signal is collected by the logic analyzer Agilent 16804A (Agilent Technologies Inc, Santa Clara, CA, USA), and then the power spectral density (PSD) analysis of the Sigma-Delta is implemented by FFT of Matlab (R2016a, MathWorks, Natick, MA, USA).

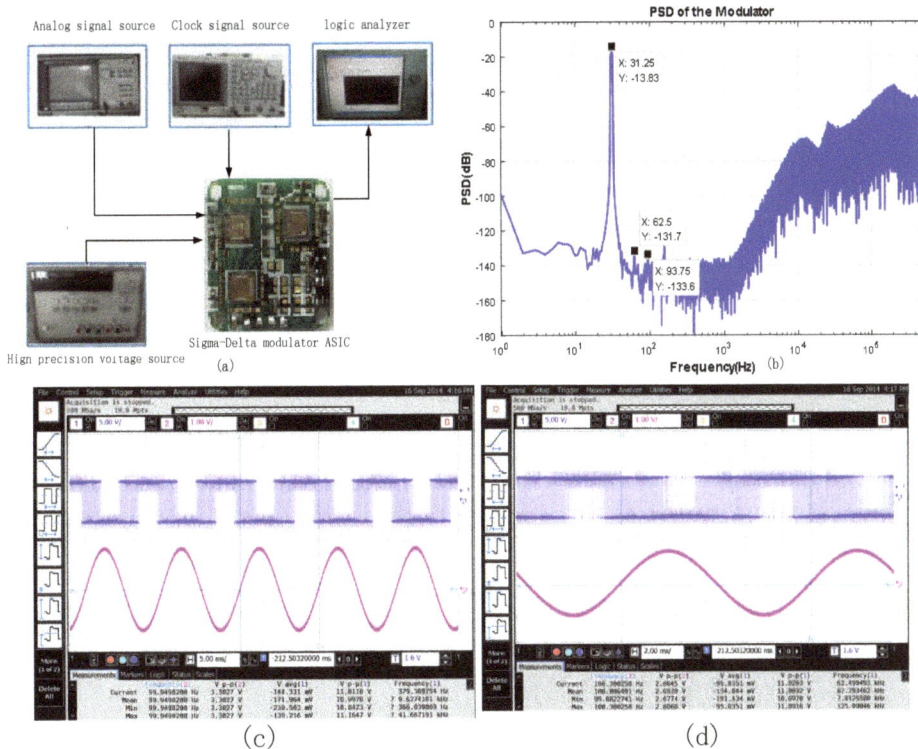

Figure 8. (**a**) The testing system of Sigma-Delta modulator Application Specific Integrated Circuit (ASIC); (**b**) The PSD of Sigma-Delta modulator; (**c**) The transient response of Sigma-Delta modulator; and, (**d**) Transient response after local amplification.

The sampling frequency is 1 MHz and the Sigma-Delta modulator bandwidth is 3.9 kHz. The input signal with a frequency of 30 Hz and an amplitude of −13 dB full scale is provided by the analog signal source. The output bit flow (65,536 points) of the designed Sigma-Delta modulator is collected by the Agilent logic analyzer 16804A (Agilent Technologies Inc, Santa Clara, CA, USA) for FFT transformation, and the output spectrum of Sigma-Delta modulator is as shown in Figure 8b. The test results show that the modulator can achieve a dynamic range of 118 dB, a SNR of 115 dB; an effective number of bits of 18.8 bits; a harmonic distortion of −110 dB in the signal bandwidth; an output background noise of −141 dB; the design of reference voltage is 2.5 V, so that the signal bandwidth equivalent input noise voltage density is 198.6 nV/Hz$^{1/2}$.

We verify the function of the modulator before testing the performance. The digital bit stream output is collected from the Sigma-Delta modulator by the oscilloscope Agilent MSO9104A (Agilent Technologies Inc, Santa Clara, CA, USA). The transient response of the modulator is as shown in Figure 8c,d. The input signal, the clock signal, and the digital output signal are in turn in the Figure 8c,d. From the transient waveform we can get that the modulator can achieve analog digital conversion function. We can verify the correctness of its function from the test results.

3.5. Design of Digital Decimation Filter

In view of the design of Sigma-Delta ADC post-stage digital filter, the three stages cascade structure (the CIC filter, CIC compensation, and finite impulse response (FIR) low-pass filter) is selected. After oversampling and noise shaping, the digital signal should be sampling down and filtered.

The sampling frequency is reduced to Nyquist frequency. The three stages cascade circuit structure is applied in the filter. The first-stage uses a cascade integral comb (CIC) filter for primary filtering and reducing to a factor of 32 desample frequency; the second-stage uses a CIC compensation filter for ripple compensation and reducing to a factor of 2 desample frequency; the third-stage uses a FIR Half-band filter for high-frequency noise filtering and reducing to a factor of 2 desample frequency. For maximum linearity, it employs a single-bit feedback DAC. The digital signal process chip STM32F405DSC (STMicroelectronicsis Company, Geneva, Switzerland) used for the temperature and linearity compensation. The data transmission to upper computer is achieved by MAX232ESE (Maxim Integrated Products, San Jose, CA, USA). The interface communication mode is SPI and RS232. The 4th-order topology structure is applied in the front-stage Sigma-Delta modulator, the first-stage CIC filter cascading number should be $L + 1$ [36–38]. The digital decimation filter model is as shown in Figure 9a. The frequency response of the first-stage CIC filter is shown in Figure 9b. From the diagram, we can see that the 5th-order cascade CIC filter can achieve a sidelobe amplitude attenuation of 70 dB and effectively suppresses the out noise of frequency band. Figure 9c shows the CIC compensation filter (decimation factor $R = 32$, differential delay factor $M = 1$, $L = 5$) cascaded band-pass amplitude response, we can be get that at the normalized frequency $w = 0.005\pi$, the amplitude can achieve -0.457 dB and -0.001 dB before and after compensation, respectively. The CIC compensation filter can flat the pass-band amplitude and make the transition zone become narrower. In order to further suppress high frequency noise, the FIR filter should have a narrow transition band. The FIR half-band filter has a sampling frequency of 15.6 kHz; a pass-band cut-off frequency of 3 kHz; a stop-band cut-off frequency of 3.5 kHz, and a pass-band attenuation of 0.001 dB. Figure 9d shows the amplitude frequency response of the half-band filter. The digital signal processing modulation is based on micro-processor software programming, so that the digital decimation filter is realized. The function verification software compiler code is transplanted to the micro-processor.

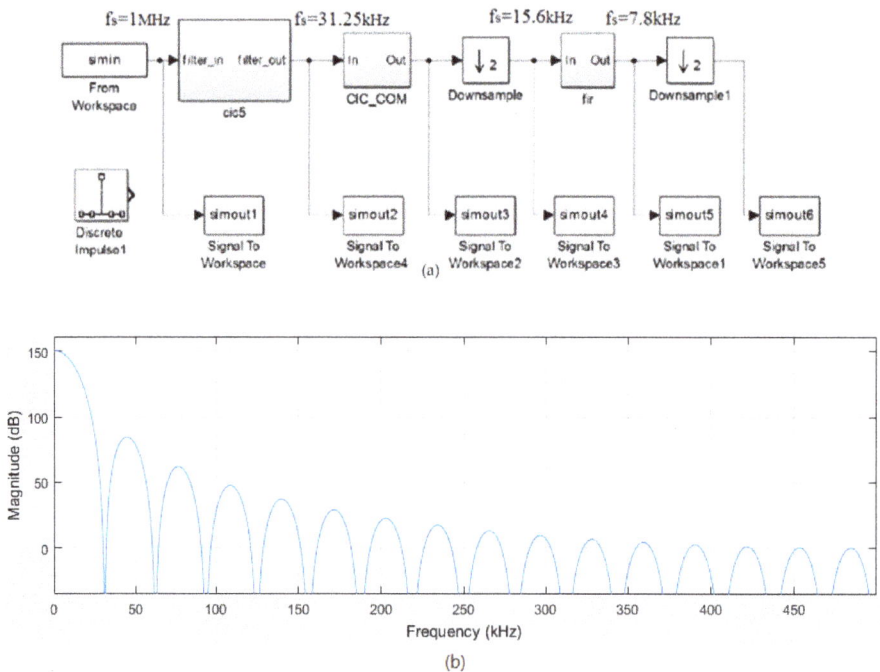

Figure 9. *Cont.*

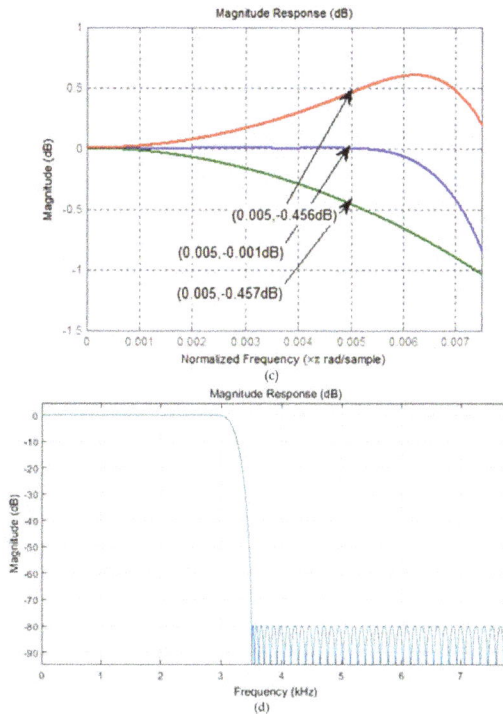

Figure 9. (**a**) Digital decimation filter model (**b**) Frequency response of first-stage cascade integral comb (CIC) filter (**c**) Passband amplitude response of CIC before and after compensation (**d**) Amplitude response of finite impulse response (FIR) low-pass filter.

3.6. The Test of Three-Axis Digital TMR Magnetometer

The test platform of the three-axis digital TMR magnetic sensor system is shown in Figure 10a. The interface circuit of the TMR sensor system was implemented in a standard 0.35 μm CMOS process. The printed circuit board (PCB) and ASIC photograph of the tunneling magneto-resistive sensor interface chip is shown in Figure 10b. The aluminum house is used for avoiding magnetic interference. The size of the digital TMR magnetic sensor system is only $25 \times 25 \times 10$ mm^3. The 5 V power supply is supported by the Agilent E3631 and the power dissipation of the TMR sensor system is 120 mW. The magnetic field is adjustable by the constant-current source Kenwood PW36-1.5ADP (Kenwood Electronics Company, Akaho, Japan). The three-axis fluxgate magnetometer FVM-400 (MEDA Company High-resolution fluxgate, Solna Municipality, Sweden) is useful for measuring the value of magnetic field. The whole test is in a three-layer shield environmental, as shown in Figure 10a. The magnetic signal is shown on personal computer (PC) by RS232 so that we can test the bias drift, linearity and noise performance of the TMR sensor system. The bias drift test of the TMR sensor system is at a zero-magnetic field environmental. The output data of the TMR sensor system is collected by PC. The TMR sensor system can achieve a bias error of 0.01% FS by standard deviation calculation.

The test of the three-axis linearity is as shown in Figure 11 by the fitting straight line at ± 1 Guass full scale (1 Guass = 10^5 nT). The test results of the three-axis linearity are x-axis 0.13% FS, y-axis 0.11% FS, z-axis 0.15% FS, respectively. The PSD of the TMR magnetometer is processed by a standard Matlab program. The TMR magnetometer system achieves a noise density of x-axis 250 pT/Hz$^{1/2}$ (at 1 Hz), y-axis 240 pT/Hz$^{1/2}$ (at 1 Hz), z-axis 240 pT/Hz$^{1/2}$ (at 1 Hz) respectively, which is limited by the low-frequency noise of the TMR sensor element.

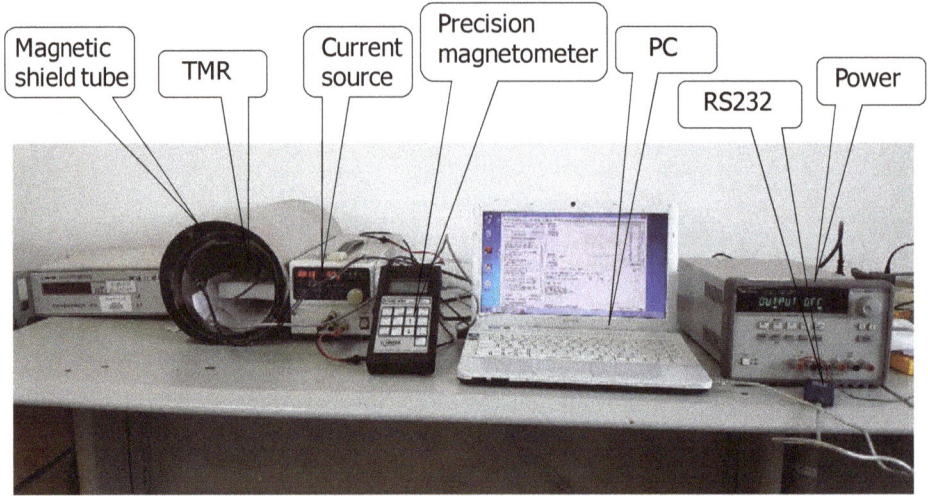

(a)

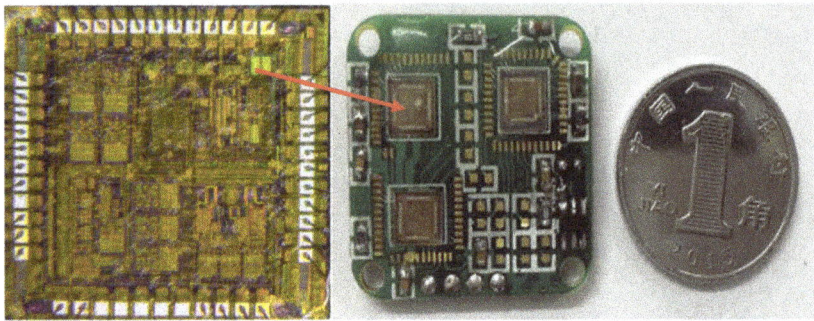

(b)

```
-2736   32693   35553   22090
-2736   32696   35553   22090
-2734   32692   35554   22091
-2736   32692   35549   22090
-2733   32692   35554   22091
-2737   32696   35554   22091
-2735   32694   35555   22091
-2737   32693   35554   22091
-2736   32694   35555   22091
-2735   32695   35554   22091
TMR SENSOR START!
DIGITAL TMR SENSOR ID = 1
TX1=115606  TX0=28982  NX3=-220  NX2=4295  NX1=18702  NX0=-249  SX=-1  ZX=-1
TY1=-21851  TY0=-24947  NY3=880  NY2=3006  NY1=24504  NY0=357  SY=-1  ZY=-1
TZ1=-34937  TZ0=-25056  NZ3=-220  NZ2=-43  NZ1=16824  NZ0=377  SZ=-1  ZZ=-1


*CHECK_FACTOR
```

(c)

Figure 10. (**a**) The test platform of the three-axis digital magnetic sensor system (**b**) The photograph of printed circuit board (PCB) and ASIC chip (**c**) The data collection by RS232 in PC.

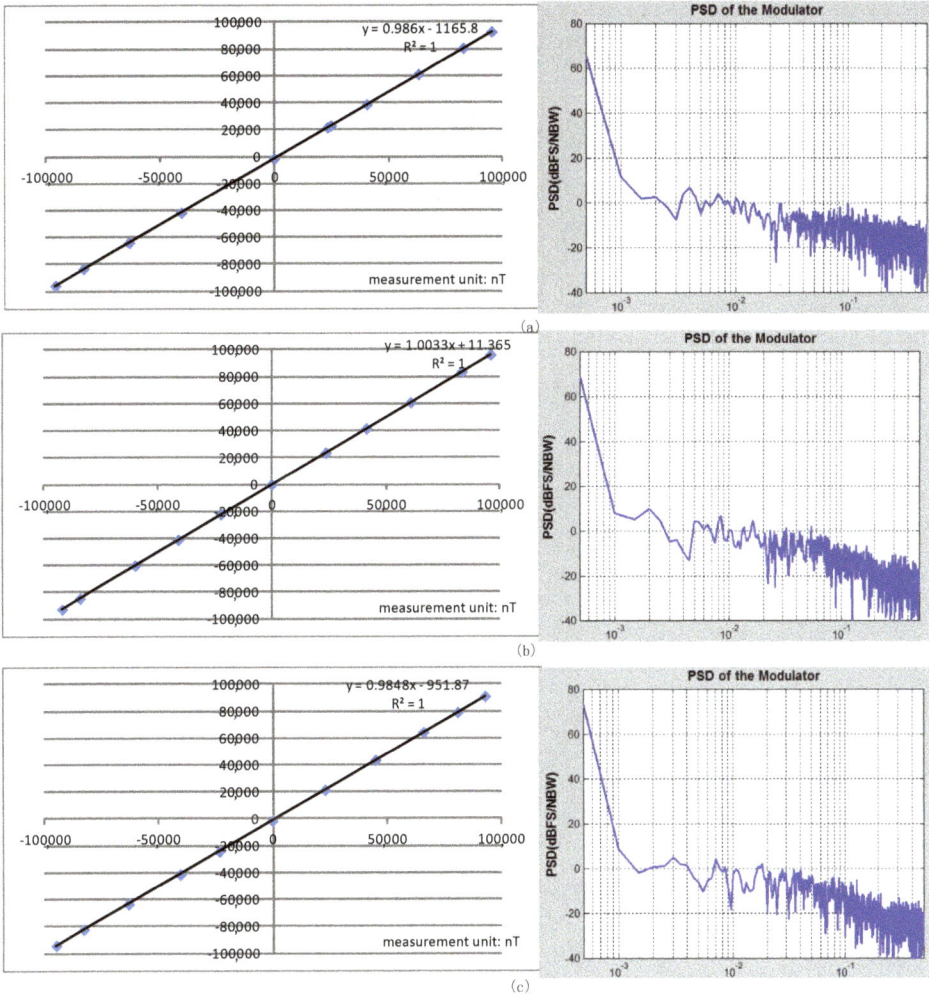

Figure 11. (**a**) The linearity and noise test of TMR magnetometer *x*-axis (**b**) The linearity and noise test of TMR magnetometer *y*-axis (**c**) The linearity and noise test of TMR magnetometer *z*-axis.

4. Conclusions

We proposed a high-precision miniaturized three-axis digital tunneling magnetic resistance-type sensor for nanosatellites' space application. The vacuum-packaged sensitive element by flux chopper can achieve a low background noise of less than 150 $pT/Hz^{1/2}$ (at 1 Hz). The interface circuit is implemented by a standard 0.35 μm 5 V CMOS process. The measurement results show that the TMR sensor system can achieve a better performance as below: at 5 V operating voltage, the sensor system can achieve a power consumption of 120 mW, a full scale of ±1 Guass, a bias error of 0.01% FS, a nonlinearity of *x*-axis 0.13% FS, *y*-axis 0.11% FS, *z*-axis 0.15% FS, and a noise density of *x*-axis 250 $pT/Hz^{1/2}$ (at 1 Hz), *y*-axis 240 $pT/Hz^{1/2}$ (at 1 Hz), *z*-axis 250 $pT/Hz^{1/2}$ (at 1 Hz), respectively. The performance of the TMR sensor element and interface circuit are both beyond Honeywell HMR2300. This magnetic sensor system satisfies the performance requirements in nanosatellites' space application.

Micromachines **2018**, *9*, 121

As shown in Table 1, this work based on the TMR sensitive structure is compared with other miniaturized magneto-resistive type magnetometers. Honeywell's HMR2300 based on the AMR sensitive structure is one of the most widely used in the U.S. GMR50 and G93 are the results of the new research and development in 2017 based on GMR sensitive structure, which have great reference value. Through the comparison of these magnetometers, the TMR magnetometer in this paper has the advantages of low noise, low power consumption and high linearity. The technical index of comprehensive performance can reach a certain level.

Table 1. Comparison of this work with other magneto-resistive magnetometers.

Miniaturized Magnetometer	Digital Output	Power/Range (V/Guass)	Consumption (mW)	Noliearity (%FS)	Noise Level (nT/Hz$^{1/2}$)	Sensitivity Axis
HMR2300	Yes	12 V/±1–2 G	400	0.1–0.5	6.67	3
GMR50	Yes	5 V/±1–4.5 G	200	0.5	10–50	1
G93	Yes	5 V/±1 G	380	0.5	10	3
This work	Yes	5 V/±1 G	120	0.11	0.25	3

Acknowledgments: The authors would like to thank National Natural Science Foundation of China (No. 61071037), Development Program of China (No. 2013AA041107) and the Fundamental Research Funds for the Central Universities (No. HIT.NSRIF.2013040) for financial support.

Author Contributions: Xiangyu Li and Jianping Hu designed the signal processing ASIC; Weiping Chen and Xiaowei Liu designed the layout of ASIC; Xiangyu Li and Liang Yin performed the experiments; and Xiangyu Li wrote the paper.

Conflicts of Interest: The authors declare no conflict of interest.

References

1. Ripka, P.; Janosek, M. Advances in Magnetic Sensors. *IEEE Sens. J.* **2010**, *10*, 1108–1116. [CrossRef]
2. Magnes, W.; Diaz-Michelena, M. Future Directions for Magnetic Sensors for Space Applications. *IEEE Trans. Magn.* **2009**, *45*, 4493–4498. [CrossRef]
3. Edelstein, A.S.; Nowak, E.R. Advances in magnetometry. *J. Phys. Condens. Matter* **2008**, *19*, 165217. [CrossRef]
4. Lenz, J.; Edelstein, S. Magnetic sensors and their applications. *IEEE Sens. J.* **2006**, *6*, 631–649. [CrossRef]
5. Freitas, P.P.; Ferreira, R.; Cardoso, S. Magnetoresistive sensors. *IEEE Transl. J. Magn.* **2007**, *19*, 165221. [CrossRef]
6. Pannetier, M.; Fermon, C.; le Goff, G. Femtotesla Magnetic Field Measurement with Magnetoresistive Sensors. *Science* **2004**, *304*, 1648–1650. [CrossRef] [PubMed]
7. Michelena, M.D.; Cerdán, M.F.; Arruego, I. NANOSAT-01: Three Years of Mission. Magnetic Scientific Results. *Sens. Lett.* **2009**, *7*, 412–415. [CrossRef]
8. Brown, P.; Beek, T.; Carr, C. Magnetoresistive magnetometer for space science applications. *Meas. Sci. Technol.* **2012**, *23*, 025902. [CrossRef]
9. Soliman, E.; Hofmann, K.; Reeg, H. Noise study of open-loop direct current-current transformer using magneto-resistance sensors. In Proceedings of the 2016 IEEE Sensors Applications Symposium (SAS), Catania, Italy, 20–22 April 2016; pp. 1–5.
10. Chinenkov, M.; Djuzhev, N.; Bespalov, V. Magnetoresistive sensor with high sensitivity: Self-aligned magnetic structures. In Proceedings of the 2017 IEEE International Magnetics Conference (INTERMAG), Dublin, Ireland, 24–28 April 2017; p. 1.
11. Lu, Y.; Deranlot, C.; Vaurès, A. Effects of a Thin Mg Layer on the Structural and Magnetoresistance Properties of CoFeB/MgO/CoFeB Magnetic Tunnel Junctions. *Appl. Phys. Lett.* **2007**, *91*, 222504. [CrossRef]
12. Lei, Z.Q.; Li, G.J.; Egelhoff, W.F. Review of noise sources in magnetic tunnel junction sensors. *IEEE Trans. Magn.* **2011**, *47*, 602–612. [CrossRef]
13. Guo, Y.; Wang, J.; White, R.M. Reduction of magnetic $1/f$ noise in miniature anisotropic magnetoresistive sensors. *Appl. Phys. Lett.* **2015**, *106*, 296. [CrossRef]

14. Pannetier-Lecoeur, M.; Fermon, C.; Vismes, A.D. Low noise magnetoresistive sensors for current measurement and compasses. *J. Magn. Magn. Mater.* **2007**, *316*, e246–e248. [CrossRef]

15. Stutzke, N.A.; Russek, S.E.; Pappas, D.P.; Tondra, M. Low-frequency noise measurements on commercial magnetoresistive magnetic field sensors. *J. Appl. Phys.* **2005**, *97*, 10Q107. [CrossRef]

16. Hu, J.; Chen, D.; Pan, M. Magnetic flux modulation with a piezoelectric silicon bridge for $1/f$ noise reduction in magnetoresistive sensors. In Proceedings of the 2013 IEEE SENSORS, Baltimore, MD, USA, 3–6 November 2013; pp. 1–4.

17. Butti, F.; Bruschi, P.; Dei, M. A compact instrumentation amplifier for MEMS thermal sensor interfacing. *Analog Integr. Circuits Signal Process.* **2012**, *72*, 585–594. [CrossRef]

18. Mohamed, A.N.; Ahmed, H.N. A low noise CMOS readout front end for MEMS BioPotential sensor applications. In Proceedings of the IEEE International Midwest Symposium on Circuits and Systems 2014, College Station, TX, USA, 3–6 August 2014; pp. 868–871.

19. Ong, G.T.; Chan, P.K. A Power-Aware Chopper-Stabilized Instrumentation Amplifier for Resistive Wheatstone Bridge Sensors. *IEEE Trans. Instrum. Meas.* **2014**, *63*, 2253–2263. [CrossRef]

20. Shen, W.; Liu, X.; Mazumdar, D. In situ detection of single micron-sized magnetic beads using magnetic tunnel junction sensors. *Appl. Phys. Lett.* **2005**, *86*, 253901. [CrossRef]

21. Hu, J.; Pan, M.; Tian, W. Integrating magnetoresistive sensors with microelectromechanical systems for noise reduction. *Appl. Phys. Lett.* **2012**, *101*, 631. [CrossRef]

22. Edelstein, A.S.; Fischer, G.A.; Pedersen, M.; Nowak, E.R.; Cheng, S.F.; Nordman, C.A. Progress toward a thousandfold reduction in $1/f$ noise in magnetic sensors using an AC microelectromechanical system flux concentrator (invited). *J. Appl. Phys.* **2006**, *99*, 08B317. [CrossRef]

23. Almeida, J.M.; Ferreira, R.; Freitas, P.P.; Langer, J.; Ocker, B.; Maass, W. $1/f$ noise in linearized low resistance MgO magnetic tunnel junctions. *J. Appl. Phys.* **2006**, *99*, 1–3. [CrossRef]

24. Fan, Q.; Huijsing, J.H.; Makinwa, K.A.A. A 21 nV/Hz$^{1/2}$ Chopper-Stabilized Multi-Path Current-Feedback Instrumentation Amplifier with 2V Offset. *IEEE Solid-State Circuits* **2010**, *47*, 464–475.

25. Li, X.; Chen, W.; Yin, L. A closed-loop Sigma-Delta modulator for a tunneling magneto-resistance sensor. *IEICE Electron. Express* **2017**, *15*, 1–6. [CrossRef]

26. Gharbiya, A.; Johns, D.A. On the Implementation of Input-Feedforward Delta–Sigma Modulators. *IEEE Trans. Circuits Syst. II Express Briefs* **2006**, *53*, 453–457. [CrossRef]

27. Michel, F.; Steyaert, M.S.J. A 250 mV 7.5 μW 61 dB SNDR SC ΔΣ modulator using near-threshold-voltage-biased inverter amplifiers in 130 nm CMOS. *IEEE J. Solid-State Circuits* **2012**, *47*, 709–721. [CrossRef]

28. Zhu, H.; Wu, X.; Yan, X. Low-Power and Hardware Efficient Decimation Filters in Sigma-Delta A/D Converters. In Proceedings of the IEEE, Conference on Electron Devices and Solid-State Circuits, Hong Kong, China, 19–21 December 2005; pp. 665–668.

29. Yu, J.; Maloberti, F. A low-power multi-bit ΣΔ modulator in 90-nm digital CMOS without DEM. *IEEE J. Solid-State Circuits* **2005**, *40*, 2428–2436.

30. Hurrah, N.N.; Jan, Z.; Bhardwaj, A. Oversampled Sigma Delta ADC decimation filter. In Proceedings of the IEEE, Design techniques, challenges, tradeoffs and optimization International Conference on Recent Advances in Engineering & Computational Sciences, Chandigarh, India, 21–22 December 2015; pp. 1–6.

31. Zhao, L.; Deng, C.; Chen, H. A 1-V 23-μW 88-dB DR Sigma-Delta ADC for high-accuracy and low-power applications. In Proceedings of the 2015 IEEE 11th International Conference on ASIC (ASICON), Chengdu, China, 3–6 November 2015.

32. Nam, K.Y.; Lee, S.M.; Su, D.K. A low-voltage low-power sigma-delta modulator for broadband analog-to-digital conversion. *IEEE J. Solid-State Circuits* **2005**, *40*, 1855–1864.

33. Zhao, Y.S.; Tang, S.K.; Ko, C.T. A chopper-stabilized high-pass Delta–Sigma Modulator with reduced chopper charge injection. *Microelectron. J.* **2011**, *42*, 733–739. [CrossRef]

34. Zhao, G.; Hu, J.; Ouyang, Y. A Novel Method for Magnetic Field Vector Measurement Based on Dual-Axial Tunneling Magnetoresistive Sensors. *IEEE Trans. Magn.* **2017**, *99*, 1–6. [CrossRef]

35. Luong, V.S.; Su, Y.H.; Lu, C.C. Planarization, Fabrication and Characterization of Three-dimensional Magnetic Field Sensors. *IEEE Trans. Nanotechnol.* **2017**, *99*, 11–25. [CrossRef]

36. Ouyang, Y.; He, J.L.; Hu, J. Prediction and Optimization of Linearity of MTJ Magnetic Sensors Based on Single-Domain Model. *IEEE Trans. Nanotechnol.* **2015**, *51*, 1–4.

Micromachines **2018**, *9*, 121

37.	Butti, F.; Piotto, M.; Bruschi, P. A Chopper Instrumentation Amplifier with Input Resistance Boosting by Means of Synchronous Dynamic Element Matching. *IEEE Trans. Circuits Syst.* **2017**, *64*, 753–764. [CrossRef]

38.	Kusuda, Y. Auto Correction Feedback for Ripple Suppression in a Chopper Amplifier. *IEEE J. Solid-State Circuits* **2010**, *45*, 1436–1445. [CrossRef]

MDPI

St. Alban-Anlage 66

4052 Basel

Switzerland

Tel. +41 61 683 77 34

Fax +41 61 302 89 18

www.mdpi.com

Micromachines Editorial Office

E-mail: micromachines@mdpi.com

www.mdpi.com/journal/micromachines

www.ingramcontent.com/pod-product-compliance
Lightning Source LLC
Chambersburg PA
CBHW041217220326
41597CB00033BA/5998